T0321113

FERMI'S PARADOX

Cosmology and Life

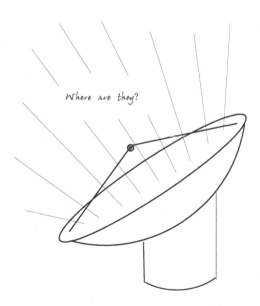

Where are they?

Michael Bodin

Order this book online at www.trafford.com
or email orders@trafford.com

Most Trafford titles are also available at major online book retailers.

Printed in the United States of America.

ISBN: 978-1-4907-4918-1 (sc)
ISBN: 978-1-4907-4919-8 (hc)
ISBN: 978-1-4907-4920-4 (e)

Library of Congress Control Number: 2014918424

Trafford rev. 10/23/2014

 www.trafford.com

North America & international
toll-free: 1 888 232 4444 (USA & Canada)
fax: 812 355 4082

This book is dedicated to my children

Judith, Richard and Pamela

Acknowledgments

My cover picture was designed and produced by Spooky Pooka Ltd, a professional artist, and I wish to express my particular thanks and appreciation for the quality and appropriateness of his work

I acknowledge my indebtedness to Wikipedia, the free encyclopedia, as an important source of general and factual information. Where appropriate, this has been duly cited, together with the Wikipedia edition from which it was obtained.

Many of the references quoted in this book can also be accessed directly through Google, or on Wikipedia (by clicking on the relevant reference itself).

I also wish to express my thanks to my daughter Pamela, and her family, for the many cups of sweet tea provided contiguously throughout the day, and which proved to be a pleasant and remarkably effective stimulant!

By the same author

Primer of Relativity - A Student's Introduction
Enigmas of Creation - Accident or Design?

The Author is a keen amateur astronomer, with interests in lunar geology, planets and solar activity, and has his own telescope and observing facilities at home.

He has a wide-ranging academic background, with degrees in physics and medicine, a PhD in neurobiology, and has published books on relativity and cosmology.

He is a past Fellow and council member of the British Interplanetary Society and Fellow of the Royal Society of Medicine. He was also UK representative on the Space Rescue committee of The International Academy of Astronautics, and is the recipient of a medal carried by the Apollo 8 astronauts on the first manned flight to the moon.

He started professional life as an RAF medical officer, and after an initial posting to a helicopter search and rescue unit, transferred to a specialist research appointment at the Royal Air Force Institute of Aviation medicine. This was the only UK facility for his particular field of interest, and also a springboard for his wider interests in space medicine, and astrobiology.

Full-time research can be narrow and restrictive, however, and having completed the work for his thesis, he decided to return to clinical practice.

Initially a Student Health medical officer at London university, he later moved to a group practice in north west London, and was also the civilian doctor for the NATO headquarter in Northwood, This was more satisfying than academic life, but the end of the era of 'family doctor', and very different from general practice today.

He now lives with his daughter in the Isle of Wight, where he divides his time between writing, astronomy and his hobbies of model shop building and collecting watches..

Contents

PART 11
LIFE

PART 111
SPECULATION

Preface - Paradoxes

"Where are they"

Enrico Fermi, Los Alamos, 1950

Rarely have three word invoked such an (inverse) response - 50 'possibilities' (at the last count); none of them an 'answer'.

Paradoxes are enigmatic, but not all enigmas are paradoxical, though they do share one thing in common - to question the validity of logic.

Yet a paradox owes much of its ambiguity to 'interpretation', and the criteria we subconsciously invoke in order to do that. We decide that criteria, in relation to what we think of as 'normal' and 'logical', and the fact that a paradox seems to conflicts with what we think is 'normal' and 'logical', means that one of us must be wrong.

Our knowledge of nature at it's most fundamental comes largely from quantum theory, and that tells us that nature pays little heed to 'logic' and 'common sense' - things behave as they do, as a whole, rather than a collection of parts.

If our understanding of nature is wrong, therefore, paradoxes may be nature's way of bringing that to our attention - they may be a consequence of misinformation, rather than a reason for it, and paradoxes may be nature's way of getting that message across.

Rather than trying to solve paradoxes, we should be looking more closely at our approach to trying to do so - how we are going about it, why we are having difficulties, why we are not succeeding.

In other words, perhaps we should be looking more at the answers than the question?

That may or may not be a logical conclusion, but it is hardly a practical approach, and we have taken the easier option, of writing a book.

It is very much an open-ended topic, however, and the huge diversity of facts and information would have been very difficult to check and confirm, without the desktop convenience of Wikipedia which made collation and presentation so much easier.

INTRODUCTION

Fermi's paradox can be stated in a number of ways, but in it's original form was reputedly only three words:

Where are they?

qualified with the surprising assertion that those who heard the comment "apparently new exactly what it refereed to".

Unfortunately, though brevity of itself can be a catchy feature, in these circumstances, it only detracts and obscures the underlying assumption on which the whole paradox depends, and which is all too often taken for granter - that extraterrestrial life does exist.

There are certainly reasoned arguments, discussed later in the book, which might tend to support that assumption, but in so far as these themselves rest on assumption (no matter how plausible) they cannot be used to justify or validate anything - least of all other assumptions.

Fermi's paradox is an example of a not uncommon situation, in which a speculative objective can achieve credibility through an extended chain of unproven associations, which as an 'entity' can all too easily be taken as 'proven fact'.

For example, the uniformity of the universe†, a necessary property for many reason, is associated with a number of principles:

- The Copernican Principle, which states that the Earth has no central or favored position within the universe (a principle which is regarded as sacrosanct).

---

† In the very widest sense of generality, that no part or aspect of the universe is in any significant way, special or different from any other.

- The Mediocrity Principle - a philosophical principle, which amounts to the statement that 'common things are commonest', and is usually regarded as essentially similar in nature to the Copernican principle.
- The cosmological Principle, which refers to the matter content of the universe, has the greatest practical significance, and states that 'the universe is homogeneous and isotropic throughout'.

In essence, while these are essentially similar in their implications, only the cosmological principle is based on firm (observational) evidence. The Copernican principle, however, tends to be accepted as 'proven fact', and in 'status' and 'common usage', is widely treated as such, though in reality no such proof exists, and correctly interpreted:

"The Copernican principle has never been proven, and in the most general sense cannot be proven, but it is implicit in many modern theories of physics".[1]

Others choose innuendo to blur the reality of an unproven supposition:

"No well-informed and rational person can imagine that the earth occupies a unique position in the universe"[2].

None of these make specific reference to 'life', however, and though Fermi's paradox was probably not relevant, the Copernican Principle was later 'upgraded' to include life, as a relativistic concept† that humans are not privileged observers of the universe.

This would not be true, however, if human life was the only life which existed, for man would then be a uniquely privileged observer; this qualification therefore can only be sustained if 'other life forms' are present elsewhere in the universe, to act as 'observers', along with human beings, who only then would no longer be privileged.

Hence we have a universe in which 'uniformity' is necessary for physical and functional reasons, duly supported by observational evidence; and as a corollary, the assumption that everything else,

---

† The general concept, initiated with respect to special relativity, that nothing can ever be absolute, but can only be judged by comparison.

including life, must be just as uniform (i.e.'common and not special'), which no matter how logical that may seem, is without factual support.

Regarding solutions to Fermi's paradox, any number of these exist, with no clear front-runner, but the reasoning behind the paradox falls into two categories:

Facts (strongly supported by findings of the Kepler space probe):

- the sun (5.7) is a typical star.
- many stars have planets (5.8).
- many of these will be earthlike.
- many will lie within the habitable zone (7.5) of their parent star.

Assumptions:

- That extraterrestrial life exists.
- Many will progress to evolve intelligence.
- Some will be at least as advanced as we are.
- These will be using radio communication much as we do.
- Stars to which these belong may have an enhanced. 'electromagnetic fingerprint', which might be visible to us.
- Some will be more advanced than we are, and some of these may have mastered interstellar travel.

The fact that the universe is so large, compared to life confined to the surface of the earth, nevertheless, does not increase the probability for the existence of life elsewhere, and nor does the suitability of other earth-like planets necessarily mean that life will develop there either.

We do not know enough about the actual origin of life, as to whether environments in which life can arise are necessarily the same as those in which it can evolve - which was not the case with the only example of life we have to go on - and we have not yet been able to explain abiogenesis (8.2), which is nevertheless the most plausible explanation (excluding Pansprermia - chapter 9).

There is also a very real possibility that life, just as for the universe itself, was created by an 'outside intelligence' - in which case, we can draw no conclusion about either from any of the knowledge we possess at the present time.

That possibility could render all our Fermi preconceptions untenable. It is also worth emphasizing that the physical evidence which does exist (fine tuning, 7.3) to support the hypothesis of 'intelligent creation' (3.8) is far stronger than any of the evidence relating to Fermi's paradox itself.

There are said to be 'at least' 50 published solutions to Fermi's question, and it is not the purpose of this book to try to distinguish between these, but rather to collate and present the information which may help readers to form their own conclusions.

The diversity of views and opinions, however, which do exist already, cover every aspect of 'life' and 'cosmology', and especially with respect to the latter, information is changing on an almost daily basis.

The book is presented in three Parts, and content is current, so far as possible, as of 12 September 2014:

Certain events subsequent to that date have been added during the course of final editing, such as the arrival of NASA's MAVEN probe into the Martian atmosphere, on 22nd September 2014, but these have not been expanded or discussed in any detail.

Part 1 covers most aspects of cosmology, with an introductory chapter on theoretical background, which summarizes relevant aspects of general physics, and the main theories on which cosmology rests today, from Newton's classical mechanics to Brane cosmology.

Part 11 is a comparable coverage of 'life;' with emphasis on the relation between that, and the universe itself.

Part 111 centers on Fermi's paradox, and is a speculative combination of facts, assumptions and opinion, with respect to the potentials for the existence of life elsewhere in the universe.

The views expressed above are important background to understanding the true nature of Fermi's paradox, which goes far beyond the literality of the three words he reputedly actually used.

It goes without saying, nevertheless, that on the basis of present knowledge, and the negative SETI evidence so far, there is no correct answer to Fermi's paradox, and in that respect, this book has tried to be neutral.

The author's final proposal (15.7,8) however, as to the eventual long term future of cosmic life, is a personal view, far too distant to be of any concern to those alive today, though their remote progeny would no doubt hold a very different view.

1. Wikipedia article (31 May 2014), Copernican Principle
2. M. Rowan-Robinson (1996), Cosmology (3rd ed.). Oxford University Press. pp. 62–63. [also quoted in Reference 1]

PART 1

THE UNIVERSE

(Chapters 1 - 5)

More complex than when Fermi posed his question.
Many more potentially habitable planets.
The potential environment for extraterrestrial life.

Chapter 1. Theoretical Background
- Introduction to the physics of cosmology.

Chapter 2. Nature of the Universe
- Overview: Origin, Properties, Destiny.

Chapter 3. Origin of the Universe
- Current Theories.

Chapter 4. Evolution of the Universe
- The Big Bang to present day.

Chapter 5. Solar System
- The habitat of known life.

Chapter 1

THEORETICAL BACKGROUND

Our knowledge of the universe dates back to antiquity, and the earliest stages of man's development - that pivotal point in evolution when 'insight and self-awareness' first differentiated human from animal life, and curiosity became the driving force towards knowledge and understanding. Yet 500 thousand years on, and those fundamental questions of origin, purpose and existence are still with us, unanswered, today.

Nevertheless, a great deal has changed, and our knowledge of the universe advanced out of all recognition, through astronomy and cosmology, to astrobiology and the possibility (some would say likelihood) that in a universe so large and complex as this, it would be surprising if earth was the only inhabited planet.

It is only within the past 300 years, however, that most of this progress has taken place, since Isaac Newton introduced universal gravitation and classical mechanics in 1687.

It was relativity and quantum theory, however, two centuries later, which entirely changed our understanding of nature, with new concepts of space and time, matter and energy, which would have been unrecognized in Newton's day.

This book covers a wide range of topics, from the nature of the universe, and the origin and evolution of organic life, to the possibilities and implications of life elsewhere, including alternative biochemistry, artificial life and artificial intelligence.

The present section is a general introduction to those physical principles and properties of nature which underpin the topics discussed in the chapters which follow.

GLOBAL THEORIES.

1.1 CLASSICAL MECHANICS:

Broadly speaking, this is the field of physics which deals with forces and motion in the everyday world in which we live, and the laws which govern their behavior. In practical terms, this means objects from the sizes of atoms upwards, and moving with speeds which remain substantially below that of light.

The main properties which characterize movement are:

- Motion: fundamental to every walk of life, and defined by Newton in terms of velocity, as the rate at which position changes, and acceleration as the rate at which velocity changes.
- Change: a non specific term for any alteration of stats quo, but it can only happen if something causes it to happen.
- Force: any influence that can bring about change.
- Work: the force (F) required to move an object through a given distance.(d), $W = F.d$.
- Momentum is a property possessed only by moving objects, and gives a measure of how difficult it would be to stop an object from moving - effectively a measure of the force necessary to bring this about. For straight line movement, liner momentum (P) is the product of (mass) x (velocity), i.e. $P = m.v$.
- Angular momentum (L) is for rotating objects, as a measure of an object's resistance (torque) required to change the velocity of

rotation; for a small object, rotating with a radius r, this is given by L = r x mv.

This has important implications in cosmology, for example, where inner regions of galaxies must rotate more rapidly than outer, to conserve angular momentum, as radius decreases.

- The Law of Conservation of momentum states that in the absence of force, momentum is always conserved, and this is an alternative way of expressing Newton's first law.

Newton defined three laws of Motion in classical mechanics:

1. A body remains at rest or uniform motion in a straight line, unless it is acted on by a force (Law of Inertia)

2. When a force (F) acts on a body of mass (m), it produces an acceleration (a) which is proportional to the force, and inversely proportional to the mass of the body: This is expressed by the equation:

$$a = \frac{F}{m}, \quad \text{Hence} \quad F = m.a$$

3. For every action there is an equal and opposite reaction.

Mass is a property of every material object, and it can be defined in two seemingly different ways:

(1) How strongly it attracts other objects (weight or gravitational mass).

(2) How difficult it is to move, or resist change in position (inertial mass).

These may not look the same, but they are identical, and both depend on the 'quantity of matter' which an object contains (equivalence principle). This highlights the fact that mass has two very different rolls, although both are related to Newton's law of universal gravitation which states:

That every particle of mass 'm', attracts every other particle of mass 'M', by a force (F), perpendicular between them, and proportional to the product of the two masses and inversely as the square of the distance (r) between them:

$$F_g = \frac{GMm}{r^2}$$

1.2 GENERAL RELATIVITY:

This theory was introduced in 1915, to extend special relativity to include gravitation, as a geometrical property of curved spacetime[1]. The central feature is also the equivalence principle[1], in which each of the following pairs of properties was identical:

- gravitational and inertial mass (as above)
- gravity and artificial acceleration
- free-fall and inertial environments

From these, Einstein drew two conclusions:

1. That gravity determines free-fall inertial frames. Hence, if we regarded space as a mosaic of many tiny frames, in each of which condition are inertial, then if we integrate these into one large single frame (the universe), we have a gravitational environment in which special relativity can apply.

2. The primary association, and corner stone of general relativity:

In the presence of gravity, the geometry of space is curved. from which Einstein made the following deductions:

- Mass gives rise to gravity, therefore in the presence of mass, space is curved.
- The path of light through space is normally a straight line; gravity curves space, therefore in the presences of gravity the path of light is curved.
- Gravity determines both how matter moves, and also the geometry (curvature) of space; therefore it is geometry, and not

'force', which determines how objects move in space. These were then combine into the following relationship:

Geometry   '''   Matter
(curvature of space)   (Contents of space)

which became the field equations of general relativity:

$$E_{iv} \quad ''' \quad \frac{8\partial G.T_{iv}}{c^4}$$

where

E_{iv} = curvature tensor,
T_{iv} = energy-momentum tensor
G   = gravitational constant,
c   = velocity of light,

The main predictions of general relativity, most of which have been confirmed, were:

- Bending of light by gravity.
- Explaining the precession of the perihelion of Mercury's orbit.
- Gravitational red shifts.
- Gravitational time dilatation.
- Gravitational waves.
- Frame dragging.

In addition to these, special relativity led to the following conclusions:

- The constant velocity of light is an absolute cosmic speed limit.
- Time and Distance reduce as velocity increases.) Insignificant
- Mass increases as velocity increases.   ) in normal life.

1.3 QUANTUM THEORY:

This deals with particles and forces within the structure of atoms, and the interaction between matter and radiation. It arose at the turn of the 20th century, as a result of difficulties with classical physics in trying to explain certain phenomena of light in terms of continuous waves.

The ultraviolet catastrophe[2] was an anomaly of black body radiation, which seemed to suggest that at thermal equilibrium this would become infinite.

The photoelectric effect[3] is another example, in which light shining onto a metal surface can cause electrons to be ejected from it. Classical physics correlates the energy of these electrons with the brightness of the beam, but in fact it was found to depend solely on frequency, and had no relation to intensity; different metals were also shown to have different threshold frequencies, below which no electrons would be ejected.

It was Einstein who proposed a solution - that instead of continuous waves, light was a stream of particles (photons), each one a packet of energy (e), proportional to it's frequency (f), where:

$$e = 1f$$

and 1 = Planck's constant.

The details of his explanation can be found in any textbook of physics, but it was revolutionary at the time, and earned Einstein his Nobel prize.

Niels Bohr introduced the first quantized model of the atom in1913, in which electrons could exist only in specified energy orbits, and if an electron jumped between orbits, one or more photons of light would be emitted, depending on the energy difference between the two orbits.

In the early twenties, Louis de Broglie proposed a theory of matter, in which particles could exhibit wave characteristics and vice versa, and wave-particle duality became an established concept - just one example of the 'weirdness' which characterizes almost every aspect of quantum physics.

The double slit experiment[4] gave elegant confirmation of this, especially when carried out using a beam of electrons - best described as 'quantum entities', since clearly we do not know what their true form is: they traveled as waves but arrive as particles.

The first mathematical description of quantum mechanics was Werner Heisenberg's matrix equations; a quantum mechanical analog of their classical counterparts, which became known as matrix mechanics. A year later Erwin Schrodinger formulated a comparable equation based on properties of waves, known as wave mechanics, and it was later shown that these two theories were identical.

The uncertainty principle, developed by Heisenberg in 1927, became the bed rock of quantum mechanics. It states that for a given pair of variables, such as position and momentum, it is not possible to determine an exact value for each at the same time; the more precisely we measure one, the less accurately can we specify the other. This is not due to any difficulty in measurement, but is an inherent property of wave-particle duality - a quantum entity simply does not possess an exact position and exact momentum at the same time. Mathematically, the uncertainty in position ($\ddot{A}_x$) and uncertainty in momentum ($\ddot{A}_p$) have a simple relationship to Planck's constant:

$$\frac{\ddot{A}_p}{\ddot{A}_x} \quad < \quad 1/2$$

Quantum uncertainty has fundamental implications for every aspect of quantum activity, and particularly with respect to properties of the vacuum (3.7) where, with a slightly more literal meaning, it regulates the random increments of time for which virtual fluctuations can exist.

Quantum probability is the other major property which underpins the quantum world. It embodies 'uncertainty', yet has no direct relation to Heisenberg's principle, nor has it any relation to conventional concepts of 'likeliness' (betting odds), or statistical averages.

Unlike classical probability, which is purely an exercise in numbers (unrelated to the events to which they refer), quantum probability is decided at the level of individual particles, and where there is a choice of outcomes, that too is effected (instantaneously) by the outcome of every other similar event going on elsewhere. The fixed result (P value) which the observer derives, then has to be compared with initial conditions, to decide what the result actually means, and there is potential subjectivity associated with doing that.

In the mid 1920s, a number of physicists collated together all of the above material to form the first comprehensive theory of quantum mechanics (Copenhagen Interpretation). This was an attempt to describe the nature of the very different 'reality' which the quantum world seemed to represent, and it embraced the following 'principles':

• Schrodinger's wave equation.
• Wave-particle duality.

- Heisenberg's uncertainty principle.
- An overall quantum probabilistic interpretation.
- That the descriptions of large scale quantum systems should approximate to classical descriptions.
- Measurements and their interpretations were classical.
- An 'outside 'observer' (? God) was necessary to make it work.

Quantum Field Theory[5] was the application of quantum principle to the electromagnetic field. A field is the area of influence surrounding an object which exerts a force, and can be described in terms of waves. In quantum theory, waves equate with particles, and that allowed the transmission of forces to be described in terms of particles: eg photons for the electromagnetic force and gravitons for gravity. These are virtual particles, but it is a matter of definition whether we regard them as 'real' or not.

The main difference between quantum mechanics and quantum field theory, is that in the former the number of particles is fixed, while in the latter they can be created or destroyed.

The Quantum Vacuum is a domain of pure energy. It is the lowest tier of physical existence, and derives power from the superposition of multiple energy fields - the electromagnet field, in all it's states, together with the those for every type of particle and interaction which exists. It has no material content, and these energy fields are manifest as 'vacuum fluctuations', forming pairs of virtual particles, and the potential interaction between these and the real world, but constrained within the increments of time set by 'uncertainty', to ensure that energy is conserved.

The ground state energy of the vacuum (true vacuum or zero- point energy) is the lowest energy level in spacetime. There are other possible energy states, however, of varying degrees of stability (false vacuum) and our universe probably exists in one of these. In quantum field theory, this sort of 'local minimum' would be metastable, but such indirect evidence as we have, suggests this would most likely be over a period of many billions of years[6].

Nevertheless, if this picture is correct, our existence could come to an end at any time, if the vacuum reverted to true. The universe would simply cease to exist. New quantum fields would generate entirely different particles and forces; and the entire structure from atoms to galaxies, would be reconstituted into a different form.

1.4 MATTER:

There is no simple way to define 'matter', in part because of the many forms which it can take, and in part because the more fundamental a definition becomes, the more the need for supplementary clarifications.

In that sense, definitions sometimes need to be arbitrary to be helpful, and for a fascinating insight into just how complex that can become, readers should refer to the 'Talk' discussion section, heading the Wikipedia article on 'Matter' (1 December 2013). Among possible definition are:

- 'Anything which is made entirely of Quarks and Leptons'.
- 'Something which takes up and occupies space; or the substance which physical objects are made of'.
- Anything possessing mass and volume, and which is interchangeable with energy, according to the equation $e = mc2$.

The material world is composed of two classes of particles - those for 'structure' (building blocks), and those which mediate 'forces' (messenger particles) - 'cement'. Their mutual relationships have a hierarchical structure:

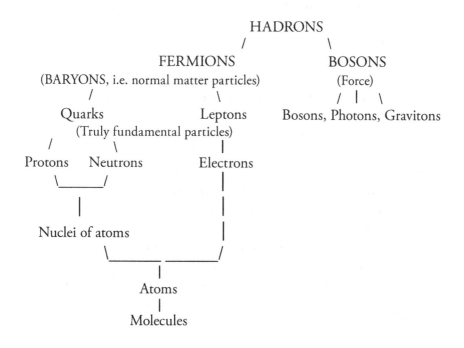

```
                        HADRONS
                     /           \
             FERMIONS             BOSONS
   (BARYONS, i.e. normal matter particles)    (Force)
         /              \              /    |    \
     Quarks           Leptons    Bosons, Photons, Gravitons
       (Truly fundamental particles)
      /      \            |
Protons   Neutrons   Electrons
    \_____/               |
       |                  |
       |                  |
  Nuclei of atoms         |
        _____   _____/
               |
             Atoms
               |
           Molecules
```

Photons and gravitons are virtual particles. They do not possess mass, though under common usage convention, they are often referred to as though they were 'real'; it is largely a matter of definition which we regarded as correct. Gravitons, however, remain hypothetical; they were introduced to explain quantum gravity, and although most versions of string theory allow for their existence, they are not supported by the Standard Model of Particle Physics. In addition, since general relativity attributes gravitation to the geometry of space, rather than 'force', gravitons can also be regarded as particles of 'curvature'.

We can now add the Higgs particle to this list; a very massive boson, formed within the first second after the Planck era, when the electroweak symmetry begins to break down, and endows other fundamental particles with their property of mass†. Under special relativity, however, mass is velocity dependent; rest mass is a fixed lower limit when measured in it's own rest frame.

Matter† can exist in one of four states: Solid, Liquid, Gas or Plasma, and in many different forms:

- Ordinary Matter (O) 4%:
  O Protons, neutrons, electrons, as
  -luminous 3.6%: Stars, luminous gases.
  -non-luminous 0.4%: Inter galactic gases
  and neutrinos.
- Dark Matter (DM) 23%: non-baryonic, unknown particles.
- Dark Energy 73%: The fundamental energy of space.

Other types of matter:

- Anti matter: anti particles - mutual annihilation on contact.
- Neutron matter: compressed matter; cores of Neutron stars.
- Nuclear matter: liquid solution of Protons and Neutrons.
- Plasma: ionized gas.
- Degenerate matter: fermion gas, at near absolute zero.
- Strange matter: liquid Quark matter.
- Exotic matter: hypothetical - matter which violates normal criteria.

---

† mass is a conserved property; matter is not.

1.5 FORCES OF NATURE:

There are four 'fundamental forces', also known as 'Fundamental interactions'. The strong nuclear force (color force) binds quarks into nucleons, and protons and neutrons into nuclei. It has a very short range, and is not active beyond the nucleus; the weak nuclear force is responsible for radioactivity (e.g. beta decay) and nuclear fusion; it is mediated by the exchange of intermediate vector bosons, which are relatively massive particles, and weak interactions essentially takes place on a point; the electromagnetic force is a 'field' force of charged electrical particles and magnetic fields, and is responsible for almost all of the phenomena in every day life: it is the interaction between the orbital electrons of adjacent atoms, which binds atoms into molecules, and is also responsible for the processes of chemical reactions.

The electromagnetic and weak interactions were later united into a single force, the electroweak interaction, and this, combined with the strong force, gave rise to the Standard Model of Particle Physics, in which all forces are now quantized,.

Gravitation, the fourth fundamental interaction, is very different from the other three. It is the weakest of the forces of nature, and this has never been properly explained; the possible influence of shadow branes[7] (1.8) is unsupported and remains pure conjecture. Gravitation belongs to classical physics, but general relativity attributes it to the curvature of space, rather than a 'force', as in it's original form. It also has a number of unique properties:

- It is a universal attractive force between any two bodies with mass.
- It's strength depend on the mass of both the attracting bodies.
- It's range is infinite, and it cannot be shielded out.
- It varies inversely with the square of distance. but effects are additive.
- It is active only on the 'large scale' and has no influence at particle level.

Attempts to unify gravity with the other three fores have been actively ongoing for over 30 years, with the development of String, Superstring and Membrane theories, but so far no insight into quantum gravity. The three quantized forces exert their actions through the exchange of gauge bosons ('messenger particles'), and to conform with that convention, a hypothetical graviton particle was introduced. This (like gravitation itself) is not supported by the Standard Model, though multi-dimensional string theory (and it's successors) do support such a particle. As an early finding, this aroused a great deal of optimism, but so far nothing more.

The 'messenger' particles for all four interactions are:

- Weak nuclear force: W and Z Bosons) Possess Mass
- Strong nuclear force: Gluons   )
- Gravitation: Gravitons   ) Do not possess
- Electromagnetic force: Photon   ) Mass

The strong nuclear force is about 100 times stronger than the electromagnetic force, but if we set the strength of the gravitational force at 1, then the relative strengths of the other three forces are[8]:

Weak / Electromagnetic / Strong - 10^{25}, 10^{36}, 10^{38}.

1.6 STANDARD MODEL:

The universe is a complex dynamic environment, and the standard model is the best theory of particle physics we have to describe it at the present time. It unites three of the four fundamental forces (strong, weak, electromagnetic), together with matter (Quarks and Leptons - all other particles are composites of these†), messenger particles (gauge bosons) which mediate the interaction of forces, and energy, into a single self consistent theory.

The theory evolved initially out of quantum electrodynamics (QED), which describes electromagnetic interactions in terms of photon exchanges between charged particles. This was later extended to incorporate the weak interaction, mediated by W and Z vector bosons, which unified the electroweak force into a single force, and thus effectively leaving only two forces to be accounted for.

---

† There are a total of approximately 18 particles in the Standard Model: 12 Fermions (6 quarks and 6 leptons), 5 gauge bosons, and the Higgs particle.

By analogy with QED, a comparable theory involving the color force - quantum chromodynamics, was later developed, in which the strong quark interaction was described in terms of gluon exchanges.

The Higgs boson has a very special role in particle physics - to explain why other fundamental particles have mass, and this can vary widely, from heavy W and Z bosons, to photons, gluons (and gravitons) which do not possess mass at all.

Confirmation of the Higgs particle (and field) in March 2013, was a powerful boost to the Standard Model, though it still falls short of a being a complete theory of fundamental interaction (theory of everything), because it is incompatible with the classical gravitation of relativity, which is not a quantized force.

Lack of quantum gravity has been a barrier to progress in physics for almost 30 years, and although usually seen as a single issue, in fact there are others (eg. time) which also need to be addressed.

1.7 STRING THEORY:

Introduced in 1980, this was a theoretical concept which replaced the point-like particles of classical physical, as fundamental constituents of matter, with tiny one-dimensional vibrating threads, no larger than the Planck length (10-33 cm).

It led to the discovery of many new 'particles', and allowed for the existence of gravitons - necessary for any theory of quantum gravity, but so far unconfirmed.

Two versions have been proposed:

Type 1 (Bosonic string theory) required 26 dimensions, many compactified, in which open-ended string increments could vibrate; there were no vibrations for fermions, however, and negative mass was permitted, which would have allowed existence of tachyons. The model lacked symmetry, never became established, and was eventually upgraded to:

Type 11 theory was the earliest precursor of superstring theory. It added vibrating closed loops to vibrating strings, and also reduced the number of dimensions from 26 to 10. In that form it embraced supersymmetry, which was found to exist between matter and massager particles, and also now including both fermions and bosons, as well as allowing for gravitons. It was also the first theory to predict shadow matter, and although an interesting possibility at the time, did not contribute significantly to further advances.

Combing Types 1 and 11, as a single theory of everything, proved almost impossible to implement, mainly on account of multiple dimensions, which could not be made compatible with the real world. To overcome this, 16 Type1 dimensions were compactified, to equal the number of Type 11 dimensions, which allowed both sets to exist in a single loop, by traveling round the loop in opposite directions. Six of these were then compactified, again by different methods, to leave only 4 dimensions, which could now be matched with those of the real world.

Nevertheless, supersymmetry remains one of the most difficult theories to understand. This was particularly true for the mathematics, and for a long time exact 'string equation' could not be derived at all, while approximations gave either multiple answers or inconsistent solutions.

By the 90s, however, that version had successfully been combined into a single theory, based on 10 space dimensions and one of time. In addition to 1-dimensional strings, a variety of other vibrating 'structures' had now been added, including 2-dimensional membranes and 3-dimensional 'bumps' (known as 'three-branes); these can be very much larger than suggested here, however, and our present universe, for example, is a 3-brane[6].

The principles did not change significantly thereafter, but all of these options eventually came to be combined into a single M- theory ('Mother' or 'Membrane' theory), which is where things stand at present. However, if 'quantum gravity' remains the criterion by which we judge success, things have advanced very little over the past three decades.

1.8 BRANE COSMOLOGY:

This looks at the relation between modern theories of cosmology and particle physics, and those of string and M-theories (as outlined above), and the possibility that the 4 dimensions of this universe may in some way be related to, or part of, a complex of higher dimensions. This would consist of large branes or even parallel universes, in the proximity of our own 3-brane domain, but not within the same spacetime framework in which that exists.

The terminology with respect to 'branes' and 'dimensions' can be confusing, but the following terms are synonyms: 'branes', 'membrane' or 'p-branes', and are defined as follows:

Spatially extended mathematical concepts in M-theory and brane cosmology, existing in a static number of dimensions[6].

The unification of nature remains a major objective in quantum physics at the present time, yet the 'Theory of Everything' is still as elusive as ever.

Nevertheless, the fact that 95% of the universe remains invisible, and it's presence only inferred through gravitational interactions with objects which we can see, means that there must be a great deal more mass (matter) in the universe than can be explained on the basis of classical physics.

There are two possibilities, however: dark matter, or variations in the strength of the gravitational force, and both of these mean that we need to review our understanding of particle physics and gravitation.

Missing mass (dark matter and dark energy) is now a dominant interest in cosmology, and that means we must think in terms of new types of matter particles. For a long time, WIMPs (Weakly Interacting Massive Particles†) have been the front runner in this respect, yet we still have no direct evidence to confirm them.

Change in the strength of gravity, on the other hand, would pose major problems for both classical physics and for relativity; 'butterfly effects' - very small changes which might be cumulative over time - can never be ruled out, but can be extremely difficult to identify and confirm.

Cosmologists are now beginning to take seriously the concept of other dimension, though 'brane' theory remains highly conjectural. The possibility of extra dimensions, however, is far from new, and Hawking, some time ago, had proposed that if large extra dimensions do exist, this would imply that our own universe would be a 4-dimensional brane, existing within some higher-dimensional domain of spacetime[6].

A further suggestion is that of parallel universe, or shadow branes, existing in the vicinity of our own brane, and might account for some of the anomalies.

The weakness of gravitation compared to the other three forces, for example, has long been a puzzle. Nevertheless, this critical relationship was responsible for significant aspects of evolution, and indeed organic life would not have been able to form at all, had the strengths of gravitation and the electromagnetic forces differed by more than a tiny amount from their present values.

---

†   Predicted by theories of particle physics, but not yet shown to exist.

There is a possible explanation for the weakness of gravitation in the fact that the graviton particle (1.4) predicted by string theory (1.7) is a closed string without ends, and as such would not be bound to the surface of branes, but able to move freely between them.'Leakage' of gravitons from our brane into surrounding higher dimensions could then account for the weakness of gravitation; and conversely, gravitons from branes adjacent to ours, might be a possible explanation for dark matter. The electroweak force, unlike gravitation, would be localized with respect to a brane, and as such would not be affected in this way..

This loss would be greater from superficial surface areas, than from deeper levels, where leakage would to some extent be shielded. It might be possible to measure this small gravitational gradient, and experiments are now in progress to see if this can be done.

Alternative dimensions (parallel universes) are very much in keeping with the concept of a multiverse (3.6), and also have important implications with respect to anthropic principles (Chapter 10). These note that life depends on a number of unlikely physical properties within this universe, and questions whether the two are related - if there were an almost infinite number of possible universes, however, this would not be so remarkable, because life would then only arise where conditions were right for it to do so.

It is unfortunate that brane cosmology is so complex, because there can be little doubt that the implications behind the theory, with respect to a much wider concept of existence than that of a single universe, may well be where some of the 'ultimate answers' will perhaps be found.

1.9 ENERGY:

This was the first property of 'reality' to arise out of the singularity at the moment of creation (2.1). Energy is interchangeable with matter, and it was this dual potential which made the universe possible. Within an instant of the Big Bang, 'singularity energy' divided into two separate pathways:

PHYSICAL EXISTENCE - spacetime, forces, particles, matter, structure.

and

ENERGY OF POWER (dark energy, cosmological constant) - Inflation, expansion, evolution.

Energy exists in a large number of different forms, and the following are among the most important:

Potential energy is energy of the status quo, and defines the inherent energy which a body or system possesses, in terms of its position (e.g. with respect to the earth's surface) before any form of change takes place, i.e. the energy which is potentially available for a specific purpose, in which work will be done.

Gravitational Potential Energy (Pg) relates to the position of an object (mass m) within a gravitational field, for example at some arbitrary height (h) above the surface of the earth, and corresponds to the work done (W) against gravity (g) to raise the object to that position i.e:

$$W = F.h$$

where the upward force (F) to move the object against the earth's gravity is equal to it's weight (mg), hence:

$$W = P_g = mgh$$

Kinetic energy, by contrast is the energy of motion. In classical mechanics, a body of mass (m) moving with velocity (v) has a kinetic energy:

$$KE = \tfrac{1}{2}mv2$$

For an object moving with a fixed velocity (v), the kinetic energy is equal to the work needed to bring it from rest up to that speed, or if it wants to stop moving, the work which it can do while slowing down to come to rest, during both of which the object would move through a distance (s) while the change in velocity was taking place.

$$\text{Work} = \text{force x distance}$$

hence:

$$W = F.d = \tfrac{1}{2}mv2$$

Kinetic energy is related to momentum (P = mv) by the equation:

$$KE = \tfrac{1}{2}mv^2 = \frac{P^2}{2m}$$

Chemical Energy is the energy involved in chemical reaction between different substances - that is, the difference between 'before' and 'after' net energy levels for each of the chemicals involved, include the energy associated with making or breaking of chemical bonds. While the reaction is taking place, energy may either be absorbed or released, depending on which of these levels is the greater.

The energy of life is a vital example, and comes from many sources, with the oxidization of food among the most important. Glucose, for example, a high energy carbohydrate, is oxidized within mitochondria to form carbon dioxide and water:

$$C_6H_{12}O_6 + 6O_2 ----- > 6CO_2 + 6H_2O$$
Glucose

and some of this energy will then be used to convert diphosphate:

$$ADP + Energy ----- > ATP + H_2O$$

Chemical energy lies at the heart of organic life, yet it embraces an evolution of critical coincidences and improbable outcomes, which make our very existence extremely unlikely. We explore the chemistry of life more fully in chapter 7.8.

NUCLEAR (binding) ENERGY is the ultimate power source of nature, and misuse a potential threat to human existence. It is the energy which went in to forming the forces which hold together protons and neutrons within the nuclei of atoms, and initially was simply part of the total energy which these individual nucleons possessed. Like all forms of energy it has mass. The mass of an intact nucleus, therefore, is always less than the sum of the masses of the individual particles, because the former no longer include the mass equivalent of the binding energy, which was used up in forming the forces which now hold the nucleons together This is also known as missing mass or 'mass defect'.

All nuclear reactions depend on the release of binding energy, according to Einstein's equation

$$e = mc^2$$

which is about a millions times greater than the energy required to ionize hydrogen atoms (electron binding energy).

In nuclear fission the nuclei of heavier elements (uranium, plutonium) are split apart; in fusion, the nuclei of light elements, (e.g. hydrogen) are combined to form the nuclei of heavier elements (e.g. helium). In both reaction, binding energy is released.

The sun and stars derive their energy from fusion, but this can only occur at very high temperatures, in order to overcome the mutual repulsion between combining protons.

1.10 TIME:

Time is the property of nature which allows for the sequential ordering of events, and ensures that everything which is about to happen, cannot all do so at once.

> "We know a great deal about time......except what it is"
> - Micho Kaku, Hyperspace[12]

Nevertheless, there are five aspects under which time can be discussed: psychological, three physical theories and various hypothetical concepts:

1. Subjective time. There is no physical equivalent of time. It does not effect any of our 5 senses, and nor do we possess other means which could make us aware of the passage of time. It is sensory input, however, which allows us to interact with our surroundings, and in turn, synchronization between that and our instinctive awareness of time, largely determines how we act and behave.

This is not conscious behavior, however, but rather a deeper subjective awareness, of something subliminal which we instinctively take for granted. If we want to know more than that, however, time suddenly becomes a mystery - essential to every aspect of our 'existence', while concealing it's own so successfully, that even today, we still do not know 'where' or 'what' time is.

It is classed as a separate dimension to distinguish it from the other three, which are physical.

2. Absolute time was introduced by Newton, as part of his 'mechanical universe' - a fixed backcloth of uniform time, the same everywhere, extending throughout the entire universe, so that measurements of time would always be the same for everyone, no matter where or how they were made. Again, it said nothing about the nature of time, but simply reflected the common sense way in which time forms part of everyday life.

3. Variable (Relativistic) time was one of the most significant predictions of special relativity, and the first concept which revealed something about it's true nature - namely, that time passes more slowly for a moving observer, than it does for an observer who remains at rest.

The 'time' itself was no different from that proposed by Newton, though the actual effect, which is related to velocity, only becomes significant at speeds approaching that of light. However, it is a true property of time, which has nothing to do with the way in which measurements are carried out, and if the speed of light ever could be reached, time would stop altogether.

This is as true for biological processes, as physical, and raises the intriguing possibility that if a spacecraft could reach sufficient speed, then for so long as that could be maintained, those on board would age very little, compared to their relatives back home (13.5). However, laws of physics make it very doubtful as to whether this could ever be carried out in practice.

4. Quantum time (5.4×10^{-44} secs) is the time required by light to travel 1 Planck length, and like all Planck units, can be defined in terms of 5 universal physical constants. It is the smallest increment of time which can exist, and also the 'age' of the universe, post singularity, at the moment of creation. The chronon is a slightly larger, though also discrete and indivisible, unit of time, with slightly different properties, and was introduced for theories attempting to unify quantum mechanics and general relativity into a viable theory of quantum gravity.

The existence (as a matter of definition rather than evidence) of such discrete increments may be suggestive, but in fact it tells us nothing about the nature of time itself at the quantum level - either it's purpose, or how it fulfills that purpose. Nor do comparisons with 'conventional' time help in this respect (where the 'arrow of time' defines it's role), because neither general relativity or the standard model of particle physics, support quantized time.

Nevertheless, there is a precedent in science for a similar paradox of physical properties, though based on proven evidence, rather than (as here) 'conflicting options' - the double slit experiment, which led to accepting 'wave-particle duality' as a property of nature, rather than trying to explain it in terms of existing science. The analogy may not be exact, but it does contrasts with the amount of time spent trying to 'quantize' the 'continuous' nature of time (and gravity) - which seem so appropriate for their actual roles in the 'macro-world'.

5. Hypotheses: The above comments suggest that time is a multi-functional property of nature, with a diversify of features, both in relation to 'what it is' and to 'how we use it', that make it difficult to treat as a single entity. The latter role in particular is largely subjective, and in that respect, it is often simpler to think of time more as a 'useful concept', than a 'physical property'.

The origin of time is one of the most debated topics in cosmology - as in Stephen Hawking's 'no boundary proposal' (3.3) for example. In general, however, there are two very different options: either time was created, and therefore must have had a beginning; or it is eternal, and has always existed.

Hawking is perhaps the strongest proponent of finite time, in which at the Planck era, immediately following the singularity, space and time were fused into one, with time, in effect, just another dimension of space. Hawking termed that a 'timeless quantum entity'. There were no pre-existing conditions, and since time did not exist, no question of a beginning.

These are not easy concepts to interpret with words, however, and they really only have true meaning expressed as equation - and even then, only to those who can understand them. Thus, to say that 'such an entity did not have a beginning, although it had not always existed, and came into being without the need for a singularity', is as meaningless to most people, as Hawking's personal assertions that pre-existing time is 'meaningless' to him.

The alternative option, however, that time has always existed, raises just as many difficulties. First, because it implies that 'something' must have been there before the universe, while at the same time effectively ruling that out, by the way in which we define what the 'universe' is; and second, what would be the purpose of time, without a universe in which to exert it's influence?

Questions of that sort are not easy to address in terms of existing science, but nevertheless, there is no reason why the physical laws, as we think we understand them at present, are necessarily comprehensive or exclusive. Time, however, is not the only area where challenging topics exist - dark matter, WIMPs, supersymmetry, branes and indeed the whole science of extra dimensions - are all areas which confront us today, and the incentives for answers which these provide, may well help to maximize the potentials of existing science, or even expand these to embrace new and unforseen areas of knowledge.

The roles of 'finite' versus 'infinite' dimensions of time are essentially similar, though the latter, in addition, has implications for creation, because it introduces the concept of pre-existence, with respect to the singularity from which the lambda-CDM model arises (3.1,4.1).

The pre-big bang ekpyrotic universe (3.4), is the only alternative proposal which actually requires pre-existence, and relates to the much wider concept of extra dimension (branes) and parallel universes. Creation in this theory occurs when two parallel branes collided.

Brane cosmology (1.8) opens up entirely new areas of physics and cosmogony, but so far purely theoretical, and at present, there is no observational evidence to support multiple dimensions, and nor do either String theory or M-theory have a need for something such as a multiverse.

Nevertheless time is arguably the most important single property of nature, not so much because of 'what it is', but because of 'how we use it' - the 'purpose' and 'functions' which it fulfills, and which make it indispensable to every aspect of 'reality' and 'existence'.

'Past', 'present' and 'future' are how we normally think of time, and this reflects one of it's most fundamental feature - the arrow of time: it 'flows' forward in one direction only, and can never reverse†. In this respect, it reflects another of natures most basic properties - entropy††,

---

†   This is true only so long as expansion continues. If that were to stop, reverse, and then start to collapse back, the arrow of time too would also change direction.

††   A measure of the disorder of a closed system, and can never decrease.

for that too can only increase, and indeed these two properties are linked with respect to one way in which the arrow of time can be defined.

Temporal divisions reflect common usage, although questions such as how we define 'present', which is the only part of time we can actually relate to, are far from straight forward. 'Past' we know to be real, because we have already lived through it, but 'future' is a closed book, though some would argue, on ground of logic alone, that the future cannot exist.

As a scale of reference, however, time comes into it's own, and fits neatly with the role of a separate dimension, additional to the three physical ones of space; but more important, it is the only property of nature which gives any meaning to physical existence, for that can only take place within an interval of time of sufficient duration for it to do so; by analogy, the reverse of quantum fluctuations (1.3, 3.7), which must cease to exist within a certain increment of time (to preserve conservation of energy).

Other forms of time19], which at some stage have been put forward as serious proposals, include: Eternalism, in which past, present and future are all present together; Presentism, in which only the present exist, and even non-existent time[9].

1.11 FUNDAMENTAL CONSTANTS:

The standard model of particle physics defines 25 dimensionless physical constants, which can be expressed only as pure numbers. They have no physical dimensions or units, and therefore have the same numerical value independently of the system of units being used - absolute properties of nature, there values unchanged throughout the physical universe.

The values of physical constants cannot be calculate, and are derived solely by measurement.

The number of such constants is not fixed, and others may be added or removed, as knowledge advances into new areas.

Groups of relative importance can vary considerable, depending on personal views, but the fine structure constant usually heads the list. Also known as the coupling constant (á) it characterizes the strength of the electromagnetic interaction, and has been accurately measured to 12 decimal places.

The question of just how constant these parameters really are is a long standing area of interest, but over the years, no confirmed variation have been found. However, some recent work in Australia, based on quasar spectra, suggest that the fine structure constant may have increased slightly over the last 104 billion years, but this has yet to be confirmed[10].

Constants of nature are of special interest to proponent of anthropic theory (Chapter 10) because of the critically narrow ranges which some of them have. Had the fine structure constant for example, been different by as little as 4%, stellar fusion would not have been able to produce carbon; or the strong nuclear force a mere 2% stronger, and the entire hydrogen content of the universe would have been consumed, within minutes of its formation by primordial nucleosynthesis, and in either case, life as it exists today would never have arisen in the first place.

There is considerable ambiguity within existing literature, regarding which 'constants' are to be regarded as 'fundamental'; the velocity of light, for example, is often listed, but is not dimensionless and therefore does not qualify.

Aside from the fine structure constant, other true dimensionless constants include,

- Strong Coupling Constant
- Number of photons per proton.
- The Charge on the electron.
- Cosmological Constant.
- Ratio of the density of the universe to the critical density.

There are also two well known subsets of dimensionless constants - those of Barrow and Tipler, in The Anthropic Cosmological Principle[11], and Martin Rees' in his book Just Six Numbers[13]: top astrophysicists, presenting detailed arguments, but nonetheless personal views, and many would disagree with them.

Reference (Chapter 1):

1. T. Hey, P. Walters, Einstein's Mirror (Cambridge University Press, 1997).
2. J. Gribben, Q is for Quantum (Weidenfeld & Nicolson, 1998)
3. A. Asimov, Asimov's Biographical Encyclopaedia of Science and Technology (Doubleday, 1964).
4. B. Green, The Elegant Universe (London: Random House, 1999).
5. J. Gribbin, Q is for Quantum: Particle Physics from A to Z (Weidenfeld & Nicolson, 1998).
6. S. Hawking, The Universe in a Nutshell (Bantum Press, 2001).
7. M. Rees, Before the Beginning (simon and Schuster Ltd, 1997).
8. Wikipedia article (24 May 2014), Fundamental Interaction (Table).
9. M. A. Bodin, Enigmas of Creation (Trafford publishing, 2010)
10. Space.com (newsletter). September 2013.
11. J.D. Barrow, J. F. Tipler, The Anthropic Cosmological Principle (Oxford University Press, 1988)
12. M. Kaku, Hyperspace (New York: Oxford University Press, 1994).
13. M. Rees, Just Six numbers (Basic Books, 1999).

Chapter 2

NATURE OF THE UNIVERSE

The Universe can be defined in a number of ways, though most tend to over-emphasise material aspects[1] at the expense of 'function' and 'evolution' (which make the universe what it is, and determine what it will become). A more balanced definition might be:

> A dynamic unbounded geometrical environment of spacetime, together with energy (matter and momentum), laws of nature, and the potential to develop life.

We believe these laws of nature to be universal and unchanging[2], for had they been otherwise, organic life would not have been able to form.

2.1 Origins (Cosmogony, chapter 3):

Early man knew little about nature. His universe was the surface of the earth and the night sky, and his only reality was life itself. Without knowledge and the science to explain it, he had no insight into 'existence', and only superstition and religion to help him understand it.

Paradoxical logic was the tool of scholars, and their ratiocinative abilities, combined with 'divine prerogative', could 'explain' origin and creation in ways that were hard to refute: St Thomas Aquinas's infinite causal regression, for example, and Aristotle's First Cause argument for the existence of God.

Historically, however, though many proposals have been advanced over the years, the absence of scientific knowledge in any true sense, inevitably limited these to the fantasies of myth and legend, which no matter how graphic, devolved ultimately to the handiwork of some unknown 'creator', and therefore effectively 'God and religion'.

There are many itemized concepts of creation with respect to the seven major religions in the world today, but for Western cultures, where the Bible remained the definitive source, there was still no attempt to explain the act itself, as we now seek to do in terms of modern science (quantum mechanics), and the biblical account simply states as 'fact', that:

"God created the universe as pleasant and pleasing....."

Genesis 1,1

with successive major features created on a daily basis over six days, ... and on the seventh day God rested.

Such ideas persisted into the Renaissance and the eventual rationalization of physical science. Astrology evolved to become Astronomy. By 1917 Vesto Slipher had established galactic redshifts and related these to recession velocity, and by the early twenties astronomers were able to explore the universe beyond the Milky Way. In 1927, Abbé Lemaître was the first to propose an expanding Universe, based on the equations of general relativity, and following a 10 year study of nearby galaxies, this was confirmed observationally by Edwin Hubble in1929.

This finding was a turning point in the history of science, which revolutionized cosmology, confirmed Einstein's equations and rationalized man's understanding of 'material existence': extrapolating backwards led to the concept of an origin at some finite time in the past, out of a hot dense state of matter, and subsequently an explanation in term of Singularity Theory[3](3.1).

It was a fortunate coincidence that relativity (1.2) came on he scene when it did, because Newton's classical mechanics was quite unsuited to the dynamic nature of the universe which was now being revealed, and in particular the concept of curved spacetime.

Einstein's equations, however, also have their limitations, for general relativity is only applicable to the 'large-scale universe' (atoms upwards), while 'creation' is concerned with the formation of energy at the lowest (sub-atomic) levels of nature, and these can only be described in terms of quantum theory.

Many different theories have been proposed over the years (chapter 3), both 'classical' and 'quantum', though so far none which have combined both theories into a single model, and following the demise of Steady-State cosmology[4] in the 1960s, the Big Bang] remained the only credible option. Most of these shared a number of properties, but the one most significant finding was, that without exception, all of them default to a common concept of 'origin' - Creation out of Nothing (3.3).

To progress further we need two things - a theory of quantum gravity, and a better understanding of the nature and role of energy itself; for what really matters at the transition from 'singularity' to

'Planck' properties, is not so much the energy, but why 'singularity energy' seems to be different from other forms. There has to be something more to it than can be accounted for just by the presence of energy alone - there needs to be 'potential', both to set in place laws of nature, and to instantly correlate these with evolution, which begins at the moment creation is over, and is continuous (not quantized) from the outset.

It may be that we can never understand such fundamental issues, but nevertheless, from that point on we know a great deal about the changes of evolution. The lambda-CDM model (4.1), in it's present form, is a good compromise theory, with many successes to it's credit, tho' some aspects of galactic formation and dark energy, remain equivocal.

Nevertheless, singularity theories will always be inherently unsatisfactory, because the initial infinites are never likely to tell us more than we know already, while to accept an 'origin from nothing' concept, will always leave doubts as to whether that really is the true position, or whether it simply marks a point beyond which the theories of physics belonging to this universe, are no longer adequate.

It is for reasons such as these that quantum alternatives are now increasingly coming into their own, with theories such as No Boundary conditions[5] (3.3), which aim to avoid the singularity, and multiple versions of String Theory (1.7), attempting, so far unsuccessfully, to develop a definitive Theory of Everything[6].

Lack of understanding quantum gravity[6] remains the main obstacle to achieving this objective, and perhaps after 40 years it is time to widen our attention to include other properties of the quantum world, such as entanglement and non-locality[7].

Nevertheless, clearly a great many issues remain to be addressed before we can expect to make real headway towards formulating a sustainable, and preferably testable, theory of how the universe began.

2.2 Expansion (Metric expansion of Space):

The space-time fabric of the universe is expanding exponentially over time. Because the Universe is defined as 'unbounded', however, with no edge in any accepted sense, it is not expanding into anything; rather, as volume enlarges with expansion, the true (proper) distance between galaxies continues to increase, while the comoving distance between them, defined by coordinates fixed with respect to expanding space itself (Hubble flow) remains unchanged.

Hubble's conclusion that the universe was expanding, based on correlating redshift measurements with recession velocity and galactic distance measurements, established two important principles in Physical cosmology: that all objects external to the Milky Way show Doppler shifts related to their velocity measured with respect the earth and to their neighbours in inter-galactic space; and that these velocities are proportional to their distance from the earth and from each other (metric expansion). We know now that these were not true Doppler effects, but rather manifestation of a cosmological red shift, due to expansion of space in which galaxies were imbedded, but for the relatively small distances involved at the time, they were indistinguishable.

Quantitatively, these results are expressed by Hubble's law:

$$v = H_0 D$$

where H0 (Hubble's constant) is the constant of proportionality between the proper (as opposed to comoving) distance of a galaxy D, and it's true velocity v, measured with respect to the earth. It is a time dependant variable and also a measure of the velocity of expansion, and the gravitational attraction of the matter content of the universe which must be overcome to avoid collapse back to a singularity†.

Despite it's importance, determining an accurate value for Hubble's constant has proved a major challenge since the outset, with some early estimates differing by almost 100%. The accepted value today, derived from studying Type Ia supernovae, lies between 70 (km/s)/Mpc and 75 (km/s)/Mpc, with a consensus towards the lower end of that range. It is a remarkably versatile parameter, and can be used to define a number of physical properties within the observable universe†, including 'time', 'age', 'size' and 'flow'†

Expansion is a consequence of the vacuum energy comprising the initial singularity, which provides the enormous repulsive force associated with dark energy and the cosmological constant. The initial inflationary phase lasted no more than 10^{-32} seconds, by which time the universe had enlarged by at least 10^{26} times, before settling back to a much slower rate. It continues to accelerate today, however, thought to be due to the influence of the cosmological constant($\ddot{E}$), which is probably dominant in the long term.

Expansion has a number of implication. The fact that it is accelerating makes it difficult to obtain accurate estimates for Hubble's constant, and indeed the universe is probably older than traditional values derived from $1/H_o$. In addition, because it is space and not matter which is expanding, that would not be subject to the constraints of special relativity, and it is quite possible that distant galaxies may now be moving apart at speeds greater than that of light, with implications for interpreting 'look back' times.

Metric expansion is the key feature of modern cosmology. It is expressed mathematically by the FLRW metric and forms the basis of the current standard (Lamda-cold-dark-matter) model, in which the geometry of the universe is highly curved on the cosmological scale. The

---

† The general symbol for Hubble's constant is H; H0 is a special case and denotes only it's value at the present time.

† That part of the total universe for which there has been time for light to reach us since the Big Bang, i.e. that part of the universe which we can see and observed.

† Hubble values include: Time/age 1/H; Length/distance (H): the distance light can travel along a linear geodesic in one Hubble time; Hubble sphere: any arbitrary volume of space-time whose radius is one Hubble length. Approximately 13.9 Gyr.

model is underpinned by three general assumptions[6], necessary to ensure uniformity across the universe, and has a high level of consistency over the whole field of observational cosmology:

1 Space is homogeneous and isotropic throughout (cosmological principle)
2 Gravitation is the only force acting.
3 Total energy is conserved.

Space can only properly be described in terms of general relativity, but we can illustrate the principle of metric expansion if we consider first a cross section through a spherical volume of uniform matter with radius substantially less than the Hubble length, in which space can be regarded as flat, and Einstein's equations reduce to those of Newton[8].

Let m be the mass of a small particle positioned on the circumference of such an area of flat space, with radius r (Figure 2.1):

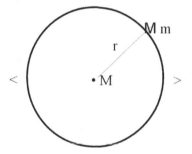

Figure 2.1
Depicting an idealized cross section of an expanding sphere of Newtonian space, with radius R < H₁

If M represents the total mass of matter within the circumference of this circular area of space, and G the gravitational constant, then the gravitational force acting radially on the particle is given by:

$$F = - \frac{GMm}{r^2} \qquad (2.1)$$

and the net gravitational force per unit mass on the particle by:

$$g = - \frac{GM}{r^2} \qquad (2.2)$$

As the circumference expands with velocity v, let the radius increase by a small amount Är over an increment of time Ät to give:

$$v = \frac{Är}{Ät} = r' \qquad (2.3)$$

This change in velocity is opposed by the centrally acting acceleration of gravity, so we can write:

$$v' = \frac{Äv}{Ät} = r'' = -\frac{GM}{r^2} \qquad (2.4)$$

To escape entirely from the gravitational influence of the disc, the particle must exceed a critical escape velocity (Ve) given by:

$$V_e = r' = \sqrt{\frac{2GM}{r^2}} \qquad (2.5)$$

The KE of expansion is ½m v^2 , or in terms of net energy per unit mass, ½v^2. If we denote this by f we can rewrite equation (2.5) in the form:

$$2f = r'^2 = \frac{2GM}{r^2} + \beta \qquad (2.6)$$

where β is a constant which will determine the eventual status of r, and can be either positive, zero, or negative, depending on whether f is greater than, equal to or less than Ve.

These equations however give a very simplistic analogy, because they apply only to the linear measures of radius vector. This can either extend, stop and remain stationery, or contract, but since the circumference of our hypothetical 'disc of space' must always match the position of the particle, in these circumstances the geometry itself can never vary.

2.3 Geometry:

It is 'dynamic' geometry however that we need to model if we are to gain any true insight into what the universe really looks like, In the above example, therefore, our 'disc of flat space' would have to be replaced by the total spherical volume of space (of which it was just a cross section), while expansion would have to be redefined in terms of comoving coordinates† by substituting the scale factor R for the numerical value of the radius r, to give a time dependant measure of change in the true physical separation between comoving galaxies as the space in which they are embedded enlarges with expansion.

A further difficulty is that we are not able to measure the total mass of the universe, but we can measure the average density of space ñ (which must be the same at all points in the universe as a requirement of the cosmological principle) with considerable accuracy, and hence derive an expression for mass:

$$M = \frac{4}{3} \pi r^3 \tilde{n}$$

Unfortunately, however, it is not possible simply to substitute these alternatives into equation 2.6, but the analogous expression with this revised terminology would be:

$$R'^2 = \frac{8}{3} \pi G R^2 \tilde{n} + \beta \qquad (2.7)$$

where â is a non specific constant, and if we replace that with a curvature constant kc^2, this now becomes the Friedmann equation for a homogeneous and isotropic expanding universe:

$$R'^2 = \frac{8}{3} \pi G R^2 \tilde{n} - kc^2 \qquad (2.8)$$

The corresponding equation for relativistic space-time is the Robertson-Walker metric, which takes the form[9]:

$$ds^2 = - dt^2 + R^2(t) \left(\frac{dr^2}{1 - kr^2} + r^2 (d\theta^2 + \sin^2\theta \, d\varphi^2) \right) \quad (2.9)$$

---

†   Comoving coordinates relate to observers who are moving with the Hubble flow, and as such are the only observers who will perceive the universe to be isotropic. The comoving distance between galaxies does not change with expansion.

where r, è and ö are comoving coordinates, and therefore remain unchanged with respect to observers expanding with the universe.

There are only three classes of geometry with homogeneous and isotropic properties. The Robertson-Walker metric applies to all three, but it is the sign of the curvature constant (0 ± 1) which determines what that will actually be - flat, spherical or hyperbolic.

Such universes are described as Friedamann or Standard universes, and the Friedmann equation (2.8) describes how the scale factor of the Robertson-Walker metric evolves over time for each of these, Figure 2.2.

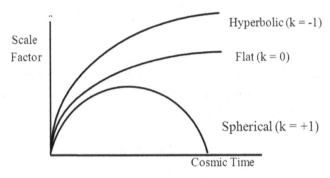

Figure 2.2 Time evolution of the expansion scale factor for the three types of Standard Friedmann universes.

A solution to equation 2.8, for given values of k and ñ is a model of the universe, and if we adjust these parameters we can compare different theoretical models with actual physical measurements.

There are two difficulties with this. First, present knowledge is based only on that part of the universe which we can see, and this may well be the smaller part of the whole; and second, once the speed of expansion, which is accelerating, exceeds that of light, no part of the universe beyond that point can ever be visible to us (ct

< D). Defining what we mean by 'shape' and 'size' therefore is not straightforward, and broadly speaking we must distinguish between local geometry, which we can measured directly, global geometry which would encompass the entire universe, largely inferred indirectly, and to what extent the former will influence the topology of the latter. Since we

cannot measure scale factors directly, however, these must be expressed in terms of parameters which we can measure.

Consider two galaxies a comoving distance d apart and true

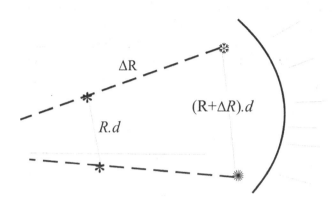

Figure 2.3

Showing increase in physical separation between two comoving galaxies as scale factor R changes with expansion.

separation R.d. where R is the scale factor. With expansion, a short interval of time Ät later and increase in scale factor ÄR, d remains unchanged, but true separation will have increased to (R + ÄR).d, and therefore velocity of expansion over that interval is given by:

$$v = \frac{\ddot{A}R.d}{\ddot{A}t} = R'.d \qquad (2.10)$$

We can also obtain the expansion velocity from Hubble's law:

$$v = H.d = H.R.d \qquad (2.11)$$

Combining these two equations we get:

$$H = \frac{R'}{R} \qquad (2.12)$$

The extent to which Hubble's parameter changes with time is related to a number of variables, including density and Friedmann's equation (which takes these into account), and describes the geometry of the universe in terms of density. The density parameter(Ω), is the ratio of the actual density(ρ), to the critical density (ρc) required for the universe to be exactly flat,

The standard Friedmann equation does not contain a cosmological constant, and in addition cosmologists have still been unable to differentiate between open and flat universes; hence, we can ignore Ë, set k = zero, and ñc then becomes equal to ñ. With these provisos we can substitute equation (2.12) into the Friedmann equation (2.8) to obtain an expression for the critical density:

$$\text{ñc} = \frac{3H^2}{8\eth G} \tag{2.13}$$

and therefore
$$\Omega = \frac{\text{ñ}}{\text{ñ}_c} = \frac{8\pi G\,\text{ñ}}{3H^2} \tag{2.14}$$

Density parameter is perhaps the most useful index we have for comparing different cosmological models, though the term as used here is the classical Ùm value which applies only to matter (dark and baryonic) density; however, in addition radiation and vacuum energy also provide density contributions in their own right, and a fuller analysis would normally take these into account.

Global geometry is complex, involving both curvature and topology, and still very much a matter of debate. Local geometry, by comparison, is straightforward, with only three classes of universe depending on whether omega is less than, equal to, or greater than 1. These values correspond to hyperbolic, flat and spherical geometries, and are illustrated in Figure 2.2. Being 4 - dimensional, however, they are not easy to visualize in Euclidian space, but we can give a fair impression of the saddle shaped pattern which lies at the centre of a hyperbolic geometry, with a simple three dimensional diagram:

Figure 2.4
Three-dimensional analog
depicting the central
portion of a
hyperbolic

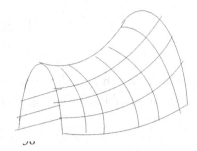

geometry.

2.4 Measurement of Distance:

As with many aspects of the universe, distance measurements are essentially hierarchical, based on a 'distance ladder' of so called 'standard candles', with each successive 'marker' calibrated against it's predecessor.

Galileo is credited with one of the earliest 'measurement' of an astronomical distance, that between the earth and the moon, which he estimated at 60 times the radius of the earth - a remarkably accurate 240,000 miles, and this was the figure later used by Newton to derive his inverse square law of gravitation.

It was the distance of stars, however, that really interested astronomers, and the earliest measurement was made by Friedrich Bessel, in 1838, when he measured the parallax of a faint star 61

Cygni, using the radius of the earth's orbit as a baseline. At 0.3 arc seconds, this was quite an achievement, and corresponded to a distance of 11 light years.

For over 70 years, this was the only method available, and beyond 300 light years (the diameter of the Milky Way is about 120,000 light years) parallax became too small to measure.

The breakthrough, which opened up the real universe, came in 1912, when Henrietta Leavitt discovered the period-luminosity relationship, which applied only to a category of stars known as Cepheid Variables. These varied in brightness over fixed periods of time, which were found to be related to their absolute magnitude (true luminosity); by comparing that with the apparent (observed) magnitude, the distance of the star could then be calculated.

This method could be used for distances up to 30 million light years, and was the basis on which Hubble compared galactic distances, in his study which confirmed the expanding universe.

Other types of variable stars, with a similar correlation between period and luminosity, later extended this up to 7 billion light years; and finally, the eventual discovery of quasars, among the brightest objects in the universe, allowed distances up to 12 billion light years to be measured i.e. within almost 10% of the Big Bang.

Hubble's work on relatively small distances, nevertheless, established the correlation between 'spectral lines shifts' and 'distance', and modern methods are now based largely on Spectroscopy.

2.5. Size and age:

These are closely related in real terms. Both are measured and expressed in similar units, and it is convenient to discuss them together. Age in particular is one of the most important properties of any universe, because it determines absolutely the suitability or otherwise for organic life to exist within it. We discus the reasons for this in chapter 7. Age and size both depend critically on the momentum of expansion imparted at the outset by the energy of the Big Bang, and on the geometry of the universe, which is decided initially by the sign of the curvature constant (k), and the value of the density parameter (Ù).

The term 'universe' is ambiguous, however, and we need to distinguish between the observable universe, which includes radiation emitted since the Big Bang (i.e. since expansion began); the visible universe which includes only radiation emitted since recombination(4.5); and the global universe which we can assume, since the discovery of accelerated expansion, is likely to be very much larger than the observable universe and possibly even infinite.

The International Astronomical Union defines 'age' as the elapsed time since the Big Bang, and current best estimates, based on measurements of cooling rate of the universe, or by extrapolating backward from the present expansion rate, place this figure at about 13.75 billion years[10]. This is light travel time however, and due to expansion of space over that interval, the proper distance of objects we see today will now be very much greater than it was when the light which we are observing actually set out. The comoving distance equivalent to that figure (which would represent the true radius of the visible universe today) is estimated to be about 46 billion light years, and therefore an observable universe with a diameter of approximately 93 billion light years. As noted earlier, for organic life to develop a universe must be old, but even that requirement would set a minimum age of only about 9,000 million light years.

2.6 Structure and Content (of Solar System, 5.4):

The universe is a four-dimensional entity of homogeneous and isotropic spacetime, finite but unbounded, which contains all forms of matter and energy, together with laws and fundamental constants of nature. It is the independent totality of everything which exists, and any suggestion of some form of alternative environment into which it could be expanding is simply meaningless.

This section is concerned only with matter content: the normal baryonic matter of everyday familiarity, which nevertheless makes up as little as 4% of the visible universe. The remaining 96% consists of dark matter and dark energy[10], which cannot be seen, and we only know they exists through indirect evidence of gravitational interactions with objects which we can see. Beyond that we know little or nothing about their true nature, though standard theories of particle physics do allow for the existence of certain particle (WIMPS - Weakly Interacting Massive Particles) which might possess some of the necessary properties, but so far remain unconfirmed.

About 3-400,000 years after the initial singularity, recombination (4.5) allowed electrons and nuclei to combined into atoms, rendering space transparent; at the same time, this allowed radiation to decouple from matter, and extend unimpeded throughout the universe. as the cosmic microwave background radiation we observe today.

Thereafter, atoms and molecules slowly condensed into clouds of matter, and after about 400 million years, into discrete entities which would later become first generation stars.

This was the beginning of an escalating process of gravitational aggregation, forming ever larger groups and groups within groups, into a hierarchical fractal structure spanning the whole extent of visible universe, and with no sign of diminishing at the present limits of observable space.

Quasars (active galactic nuclei surrounding a massive black hole) probably formed within one billion years of the initial singularity, and the earliest galaxies shortly thereafter. The nature of galaxy formation, however, is uncertain - whether smaller structures such as globular clusters accreted to form larger ones, or were a consequence of the breakup of protogalaxies.

Galactic grouping into ever larger collections of super-clusters now forms the large scale structure of the universe. These appear to remain gravitationally bound, irrespective of size, although on the grandest scale are dispersed into a looser arrangement of sheets and filaments, separated by enormous voids of relatively thinned-out space, up to 15 Mpc/h in width†1, somewhat reminiscent of the central portion of a spider's web.

The solar system may seem inconsequential by comparison with an estimated total of about 170 billion (1.7×10^{11}) galaxies within the observable universe, but from an anthropic standpoint it is more important than all of these put together. We return to this in chapter 5.

2.7 Space:

In classical terms, the greater part of the universe is empty space, but with a highly rarefied content of baryonic matter, particles and radiation, and a varying degrees of environmental dependance. We can distinguish three main categories:

Intergalactic Space approaches a perfect vacuum†2. It contains largely dark matter, dark energy and only small amounts of baryonic matter, which account for less than 4% of the total energy density approximately one proton per four cubic metres, with a mean free path of a about 10 billion light years. Nevertheless, so vast are these regions, that 'extinction effects'†3 still have to be allowed for in observational astronomy. Electromagnetic radiation and the CBR are ubiquitous. In addition, the kinetic energy of atoms crossing between galactic clusters generates enormous temperatures, sufficient to split apart electrons and nuclei, to form a hot rarefied plasma of ionized hydrogen, permeating the voids of intergalactic space.

Gravitational waves (ripples in the curvature of space-time) are a theoretical prediction of general relativity, for which there is now indirect evidence that they do exist. Such findings may seem unimportant, but they add crucial support to a growing list of marginal predictions confirming the validity and accuracy of Einstein's equations.

Interstellar space is more rarefied than the most perfect vacuum on earth, with an average density of about 105 particles per cubic metre,

---

†1 1megaparsec (Mpc) = 1 million parsecs "' 3,260,000 light years; h reflects the labile nature of Hubble constant (currently ~72 kms/s/Mps)

†2 Not to be confused with the quantum vacuum (3.7) - a hive of virtual activity, and source of vacuum energy, which powers the large scale dynamics of the universe, such as inflation and accelerated expansion.

†3 The absorption and scattering of photons by interstellar dust and gas.

rising to 115/m3 in regions of active stellar formation. Most of this consists of isolated hydrogen atoms, with smaller amounts of helium and traces of heavier elements synthesised within stars, in addition to galactic cosmic rays, other high energy particles and electromagnetic radiation. To some extent, the interstellar environment both determines the nature and activity of the objects within it, and reflects the outcome of that activity. local interstellar space is defined as that region which lies within 350 light years from the Sun, and has the potential to interact with the interplanetary environment.

Interplanetary space is largely determined by proximity to a parent star. By terrestrial standards also close to a perfect vacuum, but nevertheless a widely diversified content of plasmas, cosmic rays (ionized nuclei) and other subatomic particles, organic molecules, magnetic fields, and on a larger scale, though still extremely rarefied, clouds of dust and gas, comets and meteoric debris.

This blends smoothly with the exosphere of the terrestrial atmosphere, although the nature of such a transition might be different with respect to other planetary bodies, and could have a bearing on organic or alternative life forms which might be found to exist elsewhere.

2.8 Destiny of the Universe:

Einstein's equations of General relativity revolutionized theoretical cosmology. They explained gravity as a matter dependent consequence of curved space-time, rather than a scalar force in Euclidian space, and for the first time allowed the large scale structure of the universe to be modelled and described in terms of higher dimensional geometry.

His equations were not easy to solve, however, and could predict a large number of different universe, and each with a different long term outcome. These were explored by a number of cosmologists at the time, but Einstein himself was unhappy with his own solution, and introduced an additional term, the cosmological constant (Λ) into his equations. He was mistaken in his reasons for doing so, but since the discovery of accelerated expansion in 1998, this has reestablished itself as vacuum energy (synonymous with dark energy).

The Russian cosmologist Alexander Friedmann was one of the first to solve Einstein's equations, and the initial Friedmann universe remained the standard model for many decades.

In this universe, destiny was determined by the nature of the physical parameters established at the outset, including curvature and density parameters, and the impetus imparted by the energy density of the Big Bang. Dark matter, dark energy and the cosmological constant were not known about at the time, and the eventual fate of the universe depended only on the density of baryonic matter ($\grave{U}_m$).

If this was less than or equal to the critical density, the universe would be open and continue to expand for ever; if it were greater than $\tilde{n}_c$, expansion would slow to a halt, reverse and begin to collapse. This would be a closed universe, but the overall geometry would not be symmetrical. By comparison with the highly uniform structure of initial expansion, once the universe started to collapse, formed objects would merge ever more closely, gravitating into an irregular distribution of large discrete structures, which would eventually implode to form individual black holes. These would continue to merge, shrink and finally coalesce, to produce a single massive black hole singularity (the Big Crunch).

Like all singularities, this would not have physical existence, but is purely a mathematical concept, and we do not know what, if any, implications might have followed from it. However, though general relativity does support a Big Bang origin for our present universe, it does not allow for a bounce from a contracting phase of an earlier universe, to one of expansion[11].

Nevertheless, although considered unlikely, should the eventual fate of the universe be collapse back to a singularity, there has been interesting speculation regarding 'perception' of a terminal environment by any life forms still surviving, which might be dramatically different from that under conventional circumstances[12].

In the late 1990's the original Friedmann universe was modified to become the currently accepted lambda-CDM (ËCDM) model (4.1), which retained the FLRW metric and Friedmann's equations, and in addition incorporated cosmic inflation and vacuum energy. It still described the same three geometries, but the density parameter now included contributions from dark matter, dark energy and the cosmological constant.

The critical density is estimated to be about five atoms of hydrogen per cubic metre, while the average density of the 4.9% of baryonic matter is only about 0.2 atoms per metre Dark matter, by contrast, is estimated to account for 26.8% of the mass-energy of the universe, while dark energy which is spread extensively throughout the whole universe, makes up 68.3% of the total density. Hence, while the total mass-energy density adds up to100%, the revised density parameter (Ω) is still equal to the critical density (ñc), and we cannot distinguish (within the limits of measurement error) between open and closed universes,

However, baryonic and dark matter are both gravitationally attractive, while dark energy (the cosmological constant) is strongly repulsive, and thought to be responsible for the accelerated expansion we observe today. On that basis, therefore, a closed universe seems improbable.

More likely, expansion will continue to accelerate. Density will diminish as matter thins out; star formation will finally stop altogether, and those stars which do still exist will eventually exhaust their fuel to become white dwarfs or black holes.

Throughout all of this period, entropy would increase to a maximum, and as the universe cooled, temperatures would tend asymptotically towards absolute zero. With sufficient time, even fundamental particle, including protons, would eventually decay, leaving a universe filled only with unbound particles and radiation (heat death[11]).

References (Chapter 2):

1. J. Hawley, K. Halcomb, Foundations of Modern Cosmology (New York, Oxford University Press, 1998).
2. J. D. Barrow and F. J. Tipler, The Anthropic Cosmological Principle (Oxford University Press, 1988): 369.
3. Wikipedia article (13 September 2013), Gravitational Singularity.
4. F. Hoyle, The Intelligent Universe (Dorling Kindersley ltd, 1983.
5. J. B. Hartle, S. W. Hawking and H. Thomas, No-Boundary Measure of the Universe. Phys.Rev.Lett. 100 (2008): 201.
6. S. Hawking, The Universe in a Nutshell (London, Bantom Press, 2001).
7. M. Bodin, Enigmas of Creation (Trafford Publishing, 2010): 162.
8. J. D. Barrow and F. J. Tipler, The Anthropic Cosmological Principle (Oxford University Press, 1988): 372.
9. J. D. Barrow and F. J. Tipler, The Anthropic Cosmological Principle (Oxford University Press, 1988): 373.
10. Seven-Year Wilson Microwave Anisotropy Probe (WMAP) Observations: Sky Maps, Systematic Errors, and Basic Results (PDF). nasa.gov.
11. S. W. Hawking, The Universe in a Nutshell (Bantum Press, 2001).
12. P. Davies, The Last Three Minutes (Weidenfeld & Nicolson, 1995).

Chapter 3

ORIGIN OF THE UNIVERSE

(Cosmogony)

To clarify terminology which can be confusing, Cosmogony is concerned with the origin of the universe, Cosmology with it's evolution, nature and contents. Astronomy is the study of individual objects and the universe as a whole. The information it provides is both topical and retrospective, and so allows us to make deductions far into the past, when the universe began, and also about it's future.

(Sp ace.com)

Religion held a monopoly on questions of origin and creation, right up to modern times, when Hubble finally established the expanding universe. This was the culmination of 15 years work by a number of astronomers, studying different aspects of extra galactic movement. It was Hubble who collated the final results, though Abbe Lemaitre, three years earlier, was the first cosmologist to actually propose an expanding universe. Einstein, however, in spite of his own equations, was still reluctant to accept expansion, and it was only after a personal visit from Hubble that he finally changed his mind.

It was some time before the full implication of expansion (a finite origin at a finite time in the past) came to be generally appreciated, but they were far too radicle for the catholic church, and for a long time, Hubble's work was simply ignored.

There was some justification for this attitude, nevertheless, because no form of conventional explosive could possibly have accounted for such an explosion as the Big Bang, and this was long before the atomic bomb, or nuclear energy had been established as the power source within stars.

There was even an element of scepticism behind terms such as 'Big Bang', or the euphemism arising from Abbe Lemaitre's suggestion, in1931, of a primeval atom[1], exploding to disseminate its contents into the surrounding void, although such misleading implications certainly parallelled public thinking at the time.

Things were very different 20 years on, when a major cosmology conference was held in Rome, in 1961, by which time 'expansion' was now the established orthodoxy, and this left the Pope little choice but to finally acknowledge the importance of Hubble's work. He distanced himself from the conclusions, however, and for official church purposes, would only give the findings 'provisional' approval.

By this time cosmologists also had a much better understanding of the changes taking place in the early universe, and the enormous forces which underpinned expansion could now be attributed to the energy of a singularity.

3.1 Singularities:

The term 'Singularity' is not limited to cosmology, but applies also to a wide variety of other scientific and mathematical sub- disciplines, and is usually defined individually with respect to each of these[2].

Nevertheless, in a more general sense, it is difficult to describe singularities in a way which is really meaningful, because they have multiple characteristics, which depend very much on the context in which the term is being used. In that sense, 'cosmic singularities' are essentially a mathematical concept, representing solutions to the field equations of general relativity. As defined by the Penrose- Hawking singularity theorems[3], these can be one of two types: space-like, where matter is compressed to infinite density, e.g. by collapse of a massive star to form a black hole; or time-like, associated with regions of infinite curvature, where light rays cannot continue and come to an abrupt halt (geodesic incompleteness). So far as 'creation' is concerned, the universe is usually said to have arisen from a gravitational singularity.

However, this was not an explosion of matter in any true sense, but rather the sudden appearance of expanding space-time, arising out of a hot infinitely dense state of compressed matter, where nothing had previously existed[3]. It is even possible that there was more than one such singularity, not necessarily concurrent, which perhaps might explain the slightly uneven distribution of very distant (early) galaxies which we are beginning to see today[4]

Singularities are extremely difficult to handle, largely because solutions to the equations of general relativity most often predict a point at which the metric becomes infinite, and the whole exercise really becomes meaningless. This could just be a consequence the particular coordinates being used, but where a singularity actually does exist, it is infinite for all coordinates, irrespective of the coordinate system, and as such, permanent and cannot be removed.

It is sometimes possible to renormalize unwanted infinities, so that they effectively cancel each other out, but this can be difficult and unreliable, and cosmologists are increasingly exploring other option for creation, to try to avoid singularities altogether.

From the standpoint of 'creation', a singularity must be thought of in terms of a 'precursor' - not within the spacetime domain of this universe (as a matter of definition), but nevertheless, in some way able to become 'involved', as or when circumstance require that a new universe be brought into existence.

This hypothetical scenario would have little to do with the 'physics' of this universe, were it feasible at all. Rather, it would have to involve some process such as the self-perpetuating mechanisms of chaotic inflation (3.6), within the larger confines of a multiverse, so that the 'cause' of creation was not initially within the spacetime domain of the universe it was giving rise to.

Singularities are extremely common within our own universe, but unless we are first able to explain how they came to be formed, they cannot be used in a primary context, to account for the universe itself being created within the same spacetime as themselves.

The temperature of the universe initially at the moment of creation must have been many billions of degrees K, but this would have started to cool almost immediately, as a consequence of inflationary expansion, which enlarged the embryonic universe by up to 10^{26} times within a fraction of a second of the singularity.

3.2 Post Creation:

An early consequence of Hubble's work was to estimate the age of the universe, by backward extrapolation of expansion. Initial estimates gave values between 12 and 15 Gyrs - not too different from the currently accepted value of 13.75 billion years, but other than that, the findings posed more problems than answers - mainly relating to 'mechanisms', in the days before nuclear energy had been established as the power source within stars.

Hubble's findings were nevertheless remarkable for a number of reason: the furthest galaxies in his 'sample', for example, were a mere 30 million light years from the sun - compared with the huge comic distances deduced from them; confirming that the universe was finite settled one of the oldest argument of all time; and the incentive to explain how the universe began ushered in modern cosmology, and with it the new discipline of cosmogony.

This marked a significant change of emphasis, however, concurrent with rapid advances in instruments and technology - from 'classical' cosmology, concerned mainly with the large-scale universe that could be seen and observed, to the nature and properties of the early universe, and what could be deduced from these with respect to creation itself.

However, there were major difficulties to be overcome: the laws of Newton and Einstein applied to the physics of this universe on the large scale, while at the very 'lowest' levels of nature, where creation itself took place, the laws of quantum physics applied.

For a long time, cosmologists had been trying to unify these differing properties, and to combine them into a single comprehensive Theory of Everything, which would describe the workings of the universe at all its different levels. The one outstanding difficulty, however, was 'quantum gravity', and how to integrate that into the GUT theory of particle physics, but so far (after 30 years) without success.

A great deal of attention is rightly given to the 'physics of initiation', but cosmology covers the whole universe, and the reality is, that wether we can explain and understand the events involved in creation, has very little to do with our understanding of the rest of the universe, and it is important we keep this balance in perspective.

The current theory of the universe, the Lambda-CDM model, has an excellent explanatory correlation with the real universe from the Planck time (10^{-43} seconds) after the singularity, onwards, and we discuss the evolution and development of the adult universe in chapter 4.

3.3 No Boundary Proposal (creation out of nothing):

This is perhaps the archetypal theory of quantum cosmology, proposed by the theoretical physicists James Hartle and Stephen Hawking in 1983[5]. Arguably it remains the only truly original contribution in this field of physics since the early eighties, for although work is actively on-going in the this area, mainly in relation to supersymmetry and M-theory, the reality is that pending a breakthrough on quantum gravity, little real progress is being made.

The theory proposes that prior to the Planck era, space and time were not the separate entities we know today, but rather merged together and indistinguishable from one another, with time in effect another dimension of space The universe at that stage was a timeless quantum entity, for which the wavefunction had a transition probability that did not require any pre-existing conditions[6]. The embryonic universe in consequence had unusual properties: finite but without boundaries (analogous to the surface of a sphere), no beginning (since time did not exist), but simply appeared where nothing had previously existed. One might question how much this amounts to an explanation, but the reality is that we just do not know how the universe came to exist in the first place[7].

The theory takes its name from the fact that without prior conditions, there is no need for a singularity or for 'boundary' conditions, and that 'reality' depends only on the wave function of the universe being determined by transitions limited to 4-dimensional space, with only a single smooth finite boundary[4].

These factors are not arbitrary, however, but rather the way that this particular theory was designed, although one criticism which has been levelled, relates to the correlation between wavefunction and density of the universe, which suggests that for the Hartle and Hawking version, the initial density of such a universe might have been 'set' too low to generate the temperatures necessary to initiate inflation[6].

This work belongs to the early 80s, about the same time that interest in 'brane cosmology' began to take root, and there has been considerable overlap since then, progressing through string and superstrings to the 'Mother' theories of today, where extra- dimensions are very much becoming an issue. It can only be a matter of time, however, before theory reaches stagnation, pending the practicality of 'blending dimensions', and that in turn puts different physical laws, appropriate to the number of different dimensions (and therefore of spacetime domains) involved, on collision course. By analogy with the present position with respect to classical and quantum physics, that could be another fundamental area where 'unification' between very different set of physical laws first needs to be achieved.

It will be interesting to see what the eventual outcome with respect to quantum gravity is, but a '30 year impasse' suggests there may be something very basic standing in the way, which for example, might permanently preclude, unification between certain levels of nature.

3.4 The Ekpyrotic Universe:

This is a cosmological model based on the concept of a cyclical universe. Creation is the result of collision between two branes, and is very much a theory of 'first cause' initiation, rather than evolution. It avoids the need for a primordial singularity, but thereafter is compatible with the Big Bang standard Lambda-CDM model, and both share similar features in the very early universe.

Such a collisions would be extremely uncommon, even on a cosmological time scale, and a universe such as ours would probably be many times it's present age before the next one takes place. By the nature of such an event, however, it would not involve the physical structure of the existing universe, and presumably, the long term future too would be unaffected.

Nevertheless, these are difficult concepts to rationalize, because such branes are extra-dimensional entities. They do not belong to our present domain of spacetime, and it is unclear what sort of 'reality' would eventually result from such a collision.

It would be difficult to envisage anything of this sort, however, unless it was part of a multiverse (3.6).

3.5 Zero Energy Universe:

Hubble's discovery of expansion in the 1930s, established the need for a physical mechanism to explain creation, and also implied that a significant amount of energy must have been present, in order to power the ensuing expansion. However, without insight into the circumstances at that time, cosmologists were not in a position to speculate as to what the nature of such a mechanism might be.

Present creation theories, however, attribute this energy to a gravitational singularity (3.1), and following the discovery of the cosmic microwave background radiation a few decade later, this finally confirmed that such a large amount of energy was indeed present in the very early universe.

The instant creation was over, this energy divided, into dark energy (which is strongly repulsive) to power inflation and the ensuing expansion of the universe, which is still accelerating today; while the remainder became the gravitational (potential) energy of the matter content of the universe (which is strongly attractive). The net balance between these two opposing energies determined the nature (geometry) and long term future of the universe.

The equations of general relativity define a great many important relationships, from the equivalence of 'mass' and 'energy' ($e = mc2$) to the proportionality between 'mass' (gravitation) and 'geometry'.

Friedmann's equation is the preferred solution to Einstein's equations today, and based on that, there are only 3 possible geometries which our universe can have - open, flat and closed (Figure 2.2). When these parameters are then compared against observational findings, this confirms that the geometry of the universe is almost exactly 'flat', so far as we can tell within the limits of measurement error.

Hence, from all the available information, the two opposing forces of 'expansion' are equal, and will exactly cancel each other out, leading to the conclusion that the net total energy content of the universe must be zero.

However, there are still other sources of energy, which we have not so far taken into account.

All formed structures within this universe, from the smallest 'satellite' to the largest 'galaxy', undergo rotation, and as a consequence of that, also possess angular momentum, and the energy associated with it.

The question of whether the universe as a whole is rotating is not an easy one to answer. We simply do not know enough about the way in which it was created to form any conclusion, while direct observation (especially from 'inside') is not helpful either. In the absence of any positive confirmation, such as distortion in the uniformity of the microwave background radiation, we must conclude that the universe is not rotating; or if it is, only to an extent which would be unlikely to have any significant consequences[6].

There is one other important source of energy - electric charge, from the positive charge of protons competing with negatively charged electrons. Bearing in mind that the electroweak interaction is many orders of magnitude stronger than gravitation (the weakest of the four fundamental forces (1.5) we can take it for granted that these two forces must have exactly cancelled out, to a very high level of equality, otherwise gravitation would have been swamped, within an instant of creation, and a universe such as this would never have formed in the first place.

Hence, on the basis of the above analyses, we can now conclude that the total net balance of energy within the universe, from all possible sources, does indeed seem to be zero.

Nevertheless, that is still not a reason to conclude that a universe could, in some way, have been created out of 'nothing', for the fact that energies balance out and neutralize one another exactly, is very different from saying that energy is not present. Indeed given the amount of energy that we know went into creation, and that the universe is a totally isolated and self-contained body, clearly this is not so.

The only conclusion we can come to, therefore, from this brief analysis of 'energy', is that far from getting 'something for nothing' the most we might expect would be to obtain one zero energy universe out of another - a very different picture from that which we know prevailed at the Planck era, when this universe was created.[7] All of these arguments, however, are in the context of classical physics, which is no longer the bedrock of cosmogony, while the arguments which support Hawking's 'Creation out of Nothing'(3.3) are based on quantum interpretations of 'creation', where the laws of physics which apply at that level of nature are very different from those of classical mechanics and relativity, which apply to the rest of the universe.

Indeed, it is not just the 'physics' which is different, but the whole philosophy of expressing and interpreting results solely in terms of 'quantum probability' - one single numerical value, and which itself bears no relation to 'probability' in any 'normal' use of the term.

The whole field of physics and cosmology is now so extensive and complex, with possibilities for 'overlap' and differing 'interpretations', that if we are to make any comparison between 'classical' and 'quantum' events, no matter how similar they may seem, we must be very sure, from the outset, that we really are comparing 'like with like', before going on to draw conclusions.

3.6 Inflation:

In the Big Bang model, this is the process in which vacuum fluctuations at the end of the Planck epoch undergo exponential expansion, powered by the repulsive energy of the vacuum, which can enlarge the universe by at least 10^{26} times within 10^{-32} seconds.

It is the first step in an extended process of evolution, however, and not a mechanism for creation. As initially proposed, expansion was smooth, inconsistent with the CMB radiation, and could not account for galaxy formation. It was soon abandoned, and replaced by a slightly different form, or new inflation, which underpins most models today.

In a more general context, various inflationary scenarios have been proposed in relation to the existence of a hypothetical set of infinite universes or multiverse. Random vacuum fluctuation in any region of space in which the prevailing vacuum energy is in the false vacuum state (1.3) can undergo uneven accelerated expansion, causing fluctuations to inflate by differing amounts, to form child universes of widely varying sizes.

These remain connected to the mother universe by wormholes only, and so communication is not possible between them.

Only universes larger than about 9 billion light years are able to evolve organic life, though this would also depend on other physical parameters as well. However, although most universe would be much younger than this, other sorts of entities might exist, with a different range of properties, which might include 'artificial intelligence', and abilities to 'function' and 'perform', in ways which we would normally associate only with living beings.

It is always said that if 'alternative' universes do exist, then there is no way that we could ever know about it. However, if one looks at the sorts of predictions that are coming out of 'brane cosmology'(1.8), and the attention already being given to the possible roles of 'alternative dimensions', then even on matters as controversial as this, it is probably best to keep an open mind.

Inflation has become almost a sub-discipline of cosmology, with huge ramifications, particularly for evolution, but it is what lies behind it, and the enormity of the energies involved, which are not so easy to explain, and why it cannot be dissociated from concepts of origin.

In the most general sense, any region of space where false vacuum energy causes fluctuations to inflate, can have a variety of possible outcomes. The long term evolutionary picture can be mixed, but for simplicity we concentrate on the two which are best known, bearing in mind that in all primary and secondary inflationary scenarios, the essential mechanisms of initiation are the same for all, and it is the 'outcomes' which vary.

In chaotic inflation, large numbers of separate child universes evolve from different regions of a mother universe, as outlined above. In some of these are areas where further random fluctuations can arise spontaneously, and inflate to produce subuniverses; and these in turn will spontaneously inflate to produce more subuniverses, and so on ad infinitum. Once started, such inflationary processes cannot be stopped, and self-perpetuation will continue for ever (eternal inflation).

The question of origin becomes even more confusing in the context of whether an existing universe, which is now eternal and therefore infinite, could have had a beginning; while there are other inflating models, which no matter how large, will always be finite.

In metaphysical terms, some of the difficulties about origin and long term existence, may eventually devolve to matters of definition, although in principle, inflation is a straight forward concept, and in its basic form is now essential to explaining certain inconsistences at the outset of cosmic evolution†. There are still difficulties in reconciling inflation

---

† The cosmological problems: a number of difficulties associated with the size of the embryonic universe, which (without inflation) would have been too small to explain a number of physical characteristics of the universe today.

with quantum theory, and it also lacks confirmation in terms of particle physics[8].

3.7 Quantum Implications for Cosmogony:

These depend on properties of the quantum vacuum, which we looked at in chapter 1, and are very different from those of classical physics. It is totally devoid of matter content, and represents the lowest energy level of nature, where all fundamental quantum activity takes place - fluctuating energy fields, manifest by the formation of paired virtual particles, and the potential interaction between these and the real world.

Provided fluctuations do not exceed the increment of time allowed by uncertainty, then energy remains conserved; but if they do exceed that limit, then virtual particles become real and escape into the real world, and in doing so, deplete the vacuum by an amount of energy equal to that which was needed to form the particle in the first place.

The mechanism by which quantum activity brings about change is unclear, but what we do know is that in the quantum world, the outcome of any one event can be influenced by the outcome of all other event taking place at the same time; and each of these, in addition, will be subject to uncertainty, so it is not a simple matter of 'averages'. Indeed, one difficulty with quantum interpretations is the specificity of any result (one probability value only), which in turn requires subjective evaluation, and therefore will not necessarily mean the same for all, for 'quantum models' are neither 'actually' or 'conceptually', easy to visualize. The distinction between 'correlation' and 'explanation', can also sometimes be confusing, where the former is little more than an index of 'relevance', rather than of 'causal association', which would be implicit in the latter.

Nevertheless, quantum activity is essentially statistical (though very different from conventional), and depends on the huge numbers of individual events involved; but just how 'meaningful' these can ever be, in a literal sense, or indeed how they might translate into 'tangible physical change' in the real world, is something we do not understand.

By analogy, it is as though the cumulative effect of large numbers of events all going on concurrently, can in some way reach a consensus which can then be projected 'upwards' to influence events in the real world, and sometimes in such a way that we are unaware it has even taken place.

That may be a poor analogy, but the reality is that many of the changes we observe in the large scale universe today have their definitive origin at the quantum level of nature, and it is only comparatively recently, in historical terms, that we have come to fully realize the true potentials of subliminal activity.

In so far as quantum interpretations lack descriptive potential, and are also essentially mono-specific, they will always be of limited value, until results can be expressed in a way which is meaningful in a much wider context, and that will not necessarily be accomplished by equations alone. Nevertheless, probability will always remain a major 'index' for comparative evaluations, though how much 'importance' we place on it, may sometimes depend more on who said it than on what they said!

'No Boundaries', for example, can only really be meaningful to those able to understand the mathematics necessary to interpret the equations; while the majority (who lack such skills) must rely on a verbal descriptions, which, individually, are then usually interpreted in visual terms, but with very little true insight to correlate with Hawking's original ideas. This means two rather different interpretations in the public domain, but the importance of 'No Boundaries' is not just the proposal itself, but what that proposal represents - the archetypal quantum model which addresses 'creation' in terms of quantum (as opposed to classical) mechanics, and in doing so, obviates the need for a singularity.

From what has been said already, however, about 'creation' and 'inflation', any theory which can dispense with the energy required by the latter (and which we know was present from the outset) will have to account for a great deal more than just the act of creation itself - in fact a comprehensive alternative to classical theory, in which creation is but the first step in a complex chain of physical interactions, perpetuated indefinitely as the evolutionary future of the universe.

Confirmation of the Higgs particle and field in March 2013, marked a major advance in our understanding of matter, though the energy requirements to progress further, approaching those of the Big Bang itself, may well preclude the construction a sufficiently powerful particle accelerator within the foreseeable future†.

---

† Hawking believes that to reach such energies would require a particle accelerator larger than the diameter of the solar System[4].

Nevertheless, a number of present areas of difficulty ('dark' cosmology) are already indicating the need to review basic concepts of particle physics and matter-energy, to embrace new and unfamiliar properties which are now coming to light, and hopefully the need to simulate these will provide at least some of the incentive.

Brane cosmology which we introduced in chapter 1, opens up entirely new areas of science, to propose extra dimensions which do not even belong to the space-time domain of this universe, and as such will not be subject to laws and physical processes with which we are familiar.

The concept of 'Branes' has a theoretical background, stemming from early 'string' theories, with the eventual aim to unit GUT theory with quantum gravity. These can never be more than hypothetical, however, within the 4-dimensions of this universe, though later versions of M-theory have raised the possibility of external interactions between the spacetime domains of adjacent parallel Branes, which might for example help to explain the weakness of gravitation within our own universe (1.8).

There is one other scenario which has not so far been mentioned, and relates to a different sort of universe altogether. This is based on the statistical possibility that the life of a virtual particle, normally infinitesimal, could perhaps be very much longer - even thousands of millions of years; for 'probability' tells us that given enough time, and anything which actually is possible, must sooner or later actually come about.

Based on that, the suggestion has been made, that our whole universe might be existing as a bubble of energy inside such an inflated quantum fluctuation - a virtual 'reality' which, from the inside, we would know nothing about. The concept of an infinite multiverse adds plausibility to this, and would help to explain why, under such circumstances, we ourselves could never actually be aware of any of these other universe, because, whatever observations we make, they will only be consistent with the reality of the environment in which they are made.

Energy conservation, however, requires that the lifetime of such a universe be finite, and that makes our destiny rather messy and very insecure - instant total oblivion, at the exact moment that the 'uncertainty' initially attached to our fluctuation finally runs out (1.3).

The odds against this scenario being true must be inconceivable, but then so too are the odds against the likelihood that the universe we live in would be flat!

3.8 Divine Creation: (Higher Intelligence):

It is difficult to rationalize the diversity of features on which our existence depends, other than by accepting the reality of some underlying deeper common denominator. In cosmology, however, the distinction between science and religion is deep rooted, though 'higher intelligence' has become an acceptable euphemism for God.

We know a great deal about the universe, and about life, but nothing about why we exist, or whether there is any connection between the two; while 'fine tuning'(7.3), on the other hand, is strongly suggestive of just that - a universe created in advance for the eventuality of life, at some later stage.

That seems an acceptable answer for most people, and the logical supplementaries of by whom? and how? rarely seem to be asked, perhaps because 'God the creator' is largely self explanatory.

The early decades of the 20th century, however, established the credentials of science, for the first time, as a legitimate alternative to religion for explaining 'origin' and 'creation', by confirming the physical nature of the universe, and that it was created at a known time in the past.

This was not popular with the catholic church, however, who refused to accept the reality of Hubble's expanding universe, for almost 20 years.

Today, the church as modified it's teaching to be compatible with the real universe, while at the same time retaining it's monopoly of 'explaining creation', and given the inability of science to provide a credible alternative, looks like maintaining that position for the foreseeable future.

Nevertheless, scientists, too, have had to change and adapt, and though complex theoretical interpretations exist for what 'creation out of nothing'(3.3) really means, many still have reservations. God will always be synonymous with religion, but there is less aversion to the concept of 'outside intelligence' - entirely dissociated from the spacetime fabric of this universe, and perhaps more related to other dimensions, such as proposed by string theory(1.7) or Brane cosmology(1.8).

Nevertheless, the concept of a 'Divine Creator' is enshrined in some form or other, in almost every one of the worlds 7 major religions; while the dichotomy between 'church' and 'science' is largely a feature of Western culture, between conventional science and the Catholic church.

References (Chapter 3):

1. T. Hey and P. Walters, Einstein's Mirror (Cambridge University Press, 1997)
2. S, Hawking, A Brief History of Time (Bantu Press, 1996)
3. S. Hawking and R. Penrose, The Nature of Space and Time (Princeton University Press.1996)
4. S. Hawking, The Universe in a Nutshell (Transworld Publishers, 2001)
5. J. B. Hartle, S. W. Hawking and H. Thomas, No-Boundary Measures of the Universe, Phys.Rev.Lett, 100 (2008): 201
6. J. Barrow, The Origin Of the Universe (Weidenfeld and Nicolson, 1994)
7. L. M. Krauss, A Universe From Nothing (Kindle Edition, 2012)
8. L. Smolin, The Lifetime of the Universe (Weidenfeld & Nicolson, 1994)

Chapter 4

EVOLUTION OF THE UNIVERSE

(Cosmology)

Evolution is of central importance to cosmology, and describes the chronology of progress, by which the universe changes and develops, as structure and functions become ever more complex with age and maturity.

The initial act of creation (chapter 3), should not be seen as an isolated event, but rather as a blueprint for evolutionary continuity, whereby the initiating mechanism will embrace all the potentials necessary to mould and evolve pure energy (which is all that comes into being at the outset) into the multi-functional complex which the universe is today.

For theoreticians, designing a model universe, essentially to match 'potential' as much as 'reality', was never going to be easy, and it was not until the mid 90s that a compromise consensus was finally reached. The Lambda-CDM model which resulted now has wide general support among cosmologists, and has proved remarkably accurate with respect to most major predictions.

Quantum cosmology may well be the road ahead for those areas not yet accounted for, but it remains largely academic at present, and unlikely to gain substantive support until we are better able to understand and correlate theoretical with practical issues. Brane Cosmology (1.8 - theories based on combining particle physics, with superstrings and M-theory) has been around now since the early 80s[1], but we have to bear in mind, that while quantum theory deals with physics that we are familiar with, that is not necessarily so for the latter.

Other areas of general interest include alternatives to Einstein's gravitation, inflation and quintessence (dark energy), but these are 'topic' orientated, and there has been no serious alternative to Big Bang cosmology since the Steady State theory was ruled out.

4.1 The Lambda-CDM model (ΛCDM or standard model)
This has it's beginning in a gravitational singularity (3.1), but all evidence of that was removed by inflation, which was over by 10-32 seconds. The earliest evidence we do have comes from 9 years of WMAP data[2], and recent work with the large Hadron Collider; which can take us back to within a trillionth (10-30) of a second of creation. Provisional plans already exist for an upgraded Super Large Hadron Collider, although Hawking has stated that to probe to the energies of the Planck era, would require a particle accelerator greater than the diameter of the solar system[3].

The theory is based on general relativity and the cosmological principle†, and embraces inflation, dark matter and the cosmological constant. It's biggest drawback is that it applies only to the observable universe (which may well be only a tiny part of the cosmos as a whole), while within that, as little as 4% of matter content is actually visible to us[4], and the remaining 96% inferred from gravitational interactions with objects which we can see. Nevertheless, it accounts well for all major characteristics, including the relative abundance of light elements, the

---

† The universe has no center, and is isotropic and homogeneous throughout.

cosmic microwave background radiation, dark matter, inflation and the accelerating expansion we see today.

4.2 Evolution - the first Second:

Creation established the foundations of reality and existence, the fabric of spacetime and the laws of nature.

The first 10^{-43} seconds (Planck era) is a closed book. General relativity does not apply under such extreme conditions, and we cannot use quantum theory (which would otherwise replace it at this level of nature) without a theory of quantum gravity, which we do not yet possess[5]. Figure 4.1 illustrates initial evolutionary changes.

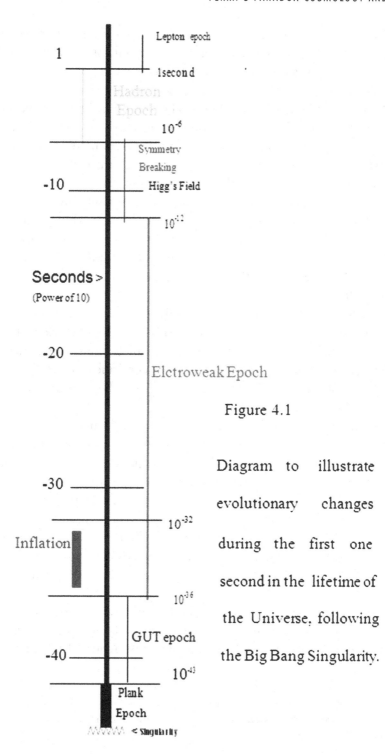

Figure 4.1

Diagram to illustrate evolutionary changes during the first one second in the lifetime of the Universe, following the Big Bang Singularity.

The 4 forces of nature† were unified on creation, but began to separate as the universe cooled with expansion. Gravity was the first to branch out, and that defined the end of the Planck epoch. The other 3 forces remained together throughout the ensuing grand unification epoch, described by the non-gravitational grand unified (GUT) theory[1] - claimed by some to be one of the most successful theories of all time.

Splitting of the strong from the electroweak force marked the end of the GUT epoch, and the onset of inflation, though there is disagreement as to wether that overlapped the electroweak epoch or had completed before it began. Inflation, since it exceeded the speed of light[6], resolved most of the cosmological problems[7] and explained the absence of monopoles in the universe today. When it was over, the universe was about the size of an orange, and the energy released filled all of space with a quark-gluon plasma, together with large numbers of exotic particles, including W and Z bosons[3].

As the temperature continued to fall, the electroweak symmetry breaks down, and one consequence of this was the formation of the Higgs field (together with the Higgs particle) which endows fundamental particles with mass. This particle was finally confirmed in March 2013, by the large Hadron Colllider[8], and professor Higgs has since been awarded a Nobel prize.

Around one second, the universe had cooled enough for matter particles (1.4) to begin to form, initially Hadrons (protons and neutrons) at 10-6 seconds, as individual quarks in the plasma combine together, and almost immediately thereafter Leptons (e.g. electrons) appeared. Both of these groups formed initially as particle-antiparticle pairs, and under normal circumstances would be expected to undergo mutual annihilation, which would bring matter formation to a halt. Fortunately, however, the eleectroweak symmetry is not absolute, and there is a very small parity imbalance, which ensured a net excess of matter over antimatter[9].

Protons and neutrons then escalated rapidly, and together with photons and electrons, formed a hot opaque plasm which persisted for the next 380,000 years, until recombination rendered the universe transparent.

---

†   Gravitation, Electromagnetism, the Weak and the Strong interactions, with the electromagnetic and weak interactions usually combined as the electroweak interaction.

4.3 Nucleosynthesis (1 - 20 minutes):

Nucleosynthesis is a general term for the physical processes which describe how the elements in the periodic table are formed. Primordial nucleosynthesis is the formation of light elements which takes place very shortly after the Big Bang, by the fusion of protons and neutrons, to form atomic nuclei. As the universe expands, temperatures fall rapidly, and there is a narrow window in the early photon epoch, which corresponds to the range of temperature compatible with nuclear fusion. This is estimated at between 10 Mev and 100 Kev, corresponding to a time interval of between <1 second and 16.6. minutes[10]. Other sources give a much shorter duration, set by the rate at which neutrons decay, with nucleosynthesis over by 3 minutes 46 seconds[11]. Irrespective of which is correct, fusion shuts down anyway after 20 minutes, when temperature and density fall below threshold with continued expansion.

In the initial reaction, protons and neutrons combine to form the hydrogen isotope deuterium:

$$P + n ----> {}^2H$$

Deuterium is unstable, and rapidly fuses into Helium-4:

$$^2H + {}^2H ----> {}^4H$$

Neutrons are unstable, with a half life of just over 10 minutes, and soon decay into protons and electrons, so that within a few minutes nucleons have become very unbalanced, with a neutron to proton ration of about 1:7, or in percentage terms 12: 84; if we redistribute these nucleons by moving 12 of the 84 protons across to join the 12 neutrons, the percentage expressed as elements becomes 24% Helium-4 to 72% Hydrogen-1, which is remarkably close to the observed proportions of 25% Helium to 75% Hydrogen in the universe today[12].

Small amounts of other elements were also formed at this time (deuterium, lithium and traces of radioactive elements) but for all practical purposes, first generation stars, were essentially composed of a pure hydrogen and helium only.

4.4 Matter and Radiation (20 minutes - 380,000 years†):

For the next 380,000 the universes was a hot dense gas of photons, electrons, helium nuclei and protons. Photons initially vastly outnumber matter particles, and the energy density of the universe was dominated by radiation. This was the radiation era, but so congested was space that the mean free path of photons was insignificant, and the environment an opaque impenetrable fog. These conditions lasted for about 70,000 years.

As the universe cooled with expansion, the energy density of photons decreased more quickly than that of matter, and this was in part because expansion stretched out their wavelengths, reducing frequency and therefore energy, until inevitably photon energy density fell below that of matter. This was a defining event, and matter domination would persist for the lifetime of the universe.

The Lambda-CDM model, however, places less emphasis on these changes, and more on cold dark matter being able to influence the effects of small perturbations on any irregularities residual from uneven inflation[15]. These might enlarge in the future, into discrete areas of lesser or greater matter density, or some might even be very early precursors of eventual large scale structures in the adult universe.

4.5 Recombination (at 380,000 years):

Up until this point in cosmic evolution, temperatures were too high for electrons to attach to protons, and baryonic matter had remained a dense opaque ionized plasma.

The universe was cooling rapidly, however, and when the temperature reached about 3-4,000K this situated began to reverse. Ions were now progressively able to capture and retain electrons, to form increasing amounts of electrically neutral hydrogen and helium atoms (recombination). This process lasted for only a few thousand years, by which time most free electrons had been combined into atoms, scattering was abolished, and the mean free path for photons soon came to exceeded the Hubble length. For the first time since creation, space had become transparent, and photons were now able to travel freely across the full extent of universe (decoupling).

These are the photons which form the microwave background radiation we observe today, and are sometimes known as relic radiation.

---

† Present theories place this upper limit slightly earlier, between 377 - 379,000years, but for convenience, the more familiar figure will be retained.

After billions of years of expansion, however, they have now cooled to a temperature of 2.7K, and their wavelength (which is inversely proportional to energy) shifted into the microwave region of the spectrum. They represent the earliest conventional (i.e. 'light') record we have of the early universe, but because they were only produced at the time of recombination (also known as the time of last scattering) photons will never be able to provide information earlier than that, i.e. the first 380,000 years in the lifetime of the universe. Fortunately, however, there are other ways in which the microwave background radiation can do this (4.6), and indeed it has proved to be not only the most incontestible evidence we have for the Big Bang itself, but also a unique means to explore the universe back to within a fraction of a second of that event.

4.6 Cosmic Microwave Background Radiation:

> The CMB is a snapshot of the oldest light in our Universe, imprinted on the sky when the Universe was just 380,000 years old. It shows tiny temperature fluctuations that correspond to regions of slightly different densities, representing the seeds of all future structure: the stars and galaxies of today..

The CMB was first predicted in 1948, with a temperature estimation of 5K, but received little attention at the time. The general question of the temperature of the universe, however, was an ongoing topic of interest, with many estimated values. In 1957, a Russian cosmologist recorded measurements which were almost certainly those of the CBR, but this was not appreciated at the time. The first generally accepted reference to detecting a microwave background radiation, again by a Russian, was only published in 1964, and ironically perhaps, within a few of months of that, Penzias and Wilson, quite by accident, made their ground breaking discovery while calibrating a prototype radio antenna. They deduce a temperature of 3.5K, and were awarded the Nobel Prize for physics in 1978.

Early attempts were made to try to explain the comic background radiation in terms of scattered star light, mainly by those who supported Steady State cosmology and disagreed with a dynamic universe. By the 1970s, however, it had been shown to have an almost perfect black body spectrum, and that could only be explained by a Big Bang.

The radiation is observed equally in all parts of the sky, and it is isotropic to a very high degree, although there is an asymmetry due to the motion of the sun (and solar system) through galactic space. This has its apex in the direction of the star Vega, so the intensity of background radiation appears greatest in that direction, and least diametrically opposite. If we plot these results as a graph of measured temperature against orientation, the curve matches a perfect cosine (Figure 4.2), but although an artifact of observation,

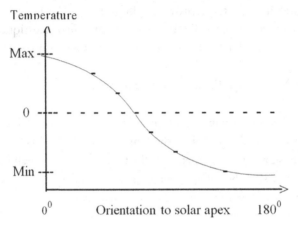

Figure 4.2

Diagram to illustrate how the measured temperature of the background radiation varies with the angular separation from the sun's path through space. ('Cosine in the sky')

it is strong confirmation that the radiation cannot have arisen locally, or it would have shared the sun's motion.

Uniformity is not 100% however, and tiny density irregularities, (anisotropy) are present, consistent with small temperature variations associated with quantum fluctuations which later underwent expansion. Some of these probably arose at the surface of last scattering (primary anisotropy), while others (secondary anisotropy) are thought to reflect thermal and gravitational interactions with ionized gasses, during their passage between then and the present day. Some anisotropies cancel out others, however, and interpretation can be difficult.

Among many studies of the microwave background, three have shown outstanding merit:

1 COBE (Cosmic Background Explore) was the first instrument designed specifically to study the CMB radiation from beyond the earth atmosphere. It was equipped with three separate instruments: a microwave radiometer, spectrometer and infrared detector, cooled with liquid helium, and the satellite was launched into space on board a Delta rocket in November 1989.

Over its 4 year life span, it was able to make full sky maps of microwave anisotropy, representing temperature fluctuations in the background radiation, to a sensitivity of one part in a 100,000. These represented density ripples in the very early universe, which were believed to be the 'seeds' of large scale structure formation, such as galactic clusters and voids which exist in the universe today. Findings were first announced in April 1992, to world wide acclaim, and Stephen Hawking described it as 'the greatest discovery of the century, if not of all time'.

Other findings, collectively from all three instruments, included 10 new galaxies, an edge-on map of our galactic disc, implications for star formation and the origin of interplanetary dust.

2 WMAP (Wilkinson Microwave Anisotropy Probe): A successor to COBE, it was 45 times more sensitive, had 33 times greater angular resolution, and it's maps contained 3,145,728 pixels. It consisted of two radiometers, attached to two Gregorian telescopes, passive thermal radiators which cooled the instruments to 90K, and a five meter solar panel base. It was launched in June 2001.

Findings have been released at two year intervals, with the 9 year data published in December 2012. Seven whole sky maps of the microwave background radiation, at frequencies ranging from 23 - 94 Ghz have been produced, and these have included 13.772 billion year old fluctuations, with a temperature range of ±200 micro-kelvin. The early universe has been shown to contain 95% dark matter and dark energy; the geometry of the universe has been shown to be within 0.4% of flat, and over the 9 years, first stars were detected a mere 200 million years after the Big Bang. The accuracy of many constants of nature were revised[2].

3 PLANCK. This satellite belongs to the European Space Agency, and was launched in May 2009. It is effectively a small space observatory, with two main instruments, which can detect the total intensity of photons over a frequence range of 30 - 857 Ghz (the CMB spectrum peaks at a frequency of 160 Ghz). The instruments are maintained at a temperature of 0.10 C above absolute zero, which makes them the coldest objects in space.

Planck can explore four additional frequency bands compared with WMAP, and also has significantly higher resolution and sensitivity, which can probe the spectrum of CMB to 3x smaller levels. In addition to supplementing WMAP however, it is equipped for a wide variety of unrelated studies, which include properties of galaxies, interstellar media, gravitational lensing, the Milky Way and the Solar System[13].

The latest data was released in March 2013, and is available in it's entirety on the internet†. This fine-tuned a number of WMAP results, providing information relating to many of these topics, and revealed the earliest ripples yet, at 10-30 seconds after the Big Bang, and thought to have given rise to a web of galactic clusters.

4.7 Dark Ages:

Recombination lasted for only a few thousand years, during which the rate of proton-electron coupling progressively exceeded that of ionization, until most of the baryonic content of the universe was bound up as neutral hydrogen.

This was the first of two phase changes which hydrogen would undergo, and it left the universe transparent, except at the 21cm waveband of neutral hydrogen (the only radiation present), and photons free to travel unimpeded. Primordial Helium undergoes similar changes, but because there is relatively less of it, and it does so at a different rate, they are of less significance.

Although the universe was no longer opaque, and photons now had unrestricted movement, in the absence of any source of light, space was dark (neutral black) and would remain that way for hundreds of millions of years, until ionization resumed once again.

The end of the dark ages began insidiously about 400 million years after the Big Bang, when structures slowly began to form, and

---

† These internet results can also be accessed directly through Wikipedia.

produce radiation able to reionze neutral hydrogen. This was the second hydrogen phase change, but as the universe enlarged and scattering was reduced, the density of space could never revert to the ionic levels before decoupling. Reionization lasted a further 400 million years, but the plasma was now one of low density hydrogen ions throughout, and space has remained transparent ever since.

There is considerable uncertainty about all aspects of the dark ages, because (by definition) there were no visible objects to study. Wikipedia gives a range from 150 - 800 million years after the Big Bang, but does not quote a source. Qasars and galaxies are said to be among the first objects formed, but only two such galaxies have actually been observed, and both of these were near the end of reionization.

These facts may be suggestive, but in so far as the dark ages overlap the very earliest stages of structure formation, they remain a topic of ongoing debate.

4.8 The Adult Universe:

There is a surprising amount of disagreement with respect to the chronology of the early universe, and structure formation in general, and for consistency we have preferred lower estimates.

Stars and galaxies began to form soon after recombination (380 million years) but the earliest star identified by WMAP was at 200 million years after the Big Bang. Most figures quoted for early stars are in the range 300 - 600 million years; galaxies appeared between 400 and 1000 million, with protogalaxies a little earlier.

Structural formation is hierarchical, and the mechanism for star formation probably constant throughout:

Interstellar space is filled with massive clouds of cold hydrogen gas, much of it in molecular form, together with small amounts of helium and cosmic dust. Areas within these clouds can begin to contract under the influence of their own gravitation, and as gas is compressed the temperature rises. Initially, this is opposed by internal gas pressure, but as the clouds slowly continue to shrink, they begin to segregate and also rotate, and this adds effects of centrifugal forces and angular momentum to an overall asymmetry.

Changes at this stage mainly affect the outer regions of the disc, and as these slowly break up, they separate into detached circumscribed areas, which can now contract independently, and eventually become small stars in their own right. Indeed most star formation results in groups or clusters of stars being formed, while isolated individual stars (such as our sun) are uncommon.

Other disc material which has not been affected in this way will either slowly scatter and disperse, or collide and clump together as planetesimals, which tend to be rocky near the center of the disc, and more gaseous towards the periphery; this protoplanetary disc will eventually segregate into a planetary system, with the same density distribution as is found in the solar system (5.2).

During this time, the main central sphere of gas continues to contract and heat up, but to create a star, the temperature must first reach a level sufficient to initiate nuclear fusion. The minimum amount of mass for this to happen is known as the initial mass function (IMF), and is about 0.08 x the mass of our sun. For significantly greater amounts of mass, other factors have to be taken into account, and to some extent the processes of star formation seem to differ with size. Very large (massive) stars however, are rare, although the majority of stars are actually smaller than our sun (5.7).

Sub-stellar objects whose mass just fails to reach the IMF level, are known as brown dwarfs; they are difficult to detect because of their faintness, but the latest estimates suggest a ratio of 1: 6 between brown dwarfs and normal stars.

As noted earlier (4.3) first generation stars are composed entirely of hydrogen and helium, and all of the heavier elements (many of which will go towards the formation organic life, for example) are produced by nuclear fusion within the cores of these stars. When they come to the end of their life, however, supernovae explosions were the dominant mechanism by which these heavier elements were then disseminated into the surrounding interstellar medium, where the next generations of stars would form.

Thus, each generation will have a slightly different composition, from the almost pure hydrogen-helium mixture of first generation stars, to progressively more metal-rich compositions in subsequent generations. These differences in composition can profoundly affect the life of a star, or may even be a deciding factor with respect to subsequent planet formation.

In general, however, when a main sequence star such as the sun exhausts it's hydrogen fuel, internal changes cause if to expand enormously. This will be the eventual fate of our sun in about 5 billion years, when it will enlarge as far out as the orbit of the earth, to become a red giant; in the very long term, it's core will eventually contract to form a white dwarf.

Quasars, thought to be galactic nuclei surrounding massive central black holes, have presented many puzzles. Now accepted as among the oldest objects in the universe, their initial identification was delayed by almost 2 years, because their spectra were so red shifted (due to distance) that for a long time they could not be recognized. Nevertheless, the high iron levels in the spectra of some quasars could only have come from supernovae[14]. This meant that these particular quasars could not have formed until after first generation stars had completed their life cycle, and that would place their age at least around 1,000 million years post Big Bang.

Figure 4.3 summarises the major evolutionary changes over the lifetime of the universe, which is now put at 13.787 Gyrs, based on the latest WMAP data (December 2012).

Structural evolution thereafter proceeded hierarchically, with smaller objects aggregating to former larger ones, and there are five levels of structure larger than galaxies - groups, clusters, clouds, super clusters and super cluster complexes or walls, which enclose

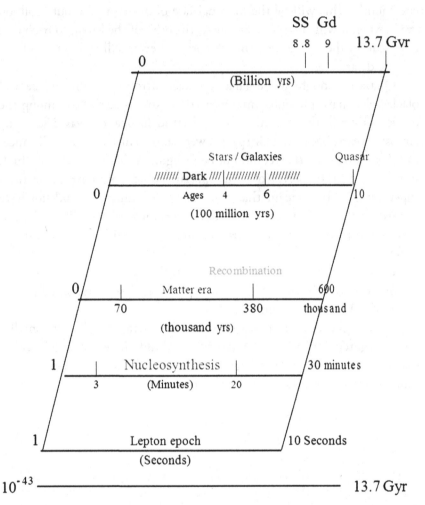

Figure 4.3
Diagram to illiterate the temporal evolution of the universe, from one
second after the Big Bang singularity, to the present day (not to scale).
[Note individual time scales for each level of evolution]

SS = Solar System formation
Gd = Galactic Disc of Milky Way formed.

voids (cosmic bubbles). No limit to clustering has yet been found within
the observable universe, and the ultimate pattern superimpose on to that
is thought to be fractal,

References (Chapter 4):

1. S. Hawking, A Brief History of Time (Bantam Press, 1996).
2. Wikipedia article (28 September 2013), Wilkinson Microwave Anisotropy Probe.
3. S. Hawking, The universe in a Nutshell (Bantum Press, 2001)
4. M. Reese, Before the Beginning (Simon & Schuster Ltd, 1997)
5. J. D. Barrow & F. J. Tipler, The Anthropic Cosmological Principle (Oxford University Press, 1994).
6. T. Ferris, The Whole Shebang, (Weidenfeld & Nicolson, 1997)
7. J. Hawley and K. Holcomb, Foundations of Modern Cosmology (Oxford University Press, 1998).
8. Wikipedia article (22 January 2014), Higgs Boson.
9. S. Hawking, The Universe in a Nutshell (Transworld Publishers, 2001).
10. Wikipedia article (15 December 2013), Big Bang nucleosynthesis (Characteristics).
11. J. Gribben, Q is for Quantum (Weidenfeld & Nicolson, 1998).
12. Big Bang Cosmology, Phys. Rev. D86, 010001 (2012).
13. Wikipedia article (11 March 2014), Planck (spacecraft)
14. T. Ferris, The Whole Shebang (Weidenfeld & Nicolson, 1997).

Chapter 5

THE SOLAR SYSTEM

Early astronomical records were essentially general by nature, based entirely on naked-eye observations. Trying to interpret these within the larger context of a universe, in which the earth was clearly central, though it was man and Gods who really mattered, was a difficult task, and usually undertaken by the philosopher- scholars of the day.

5.1 Background:
Early descriptions of celestial objects are as old as records themselves, but what is not so clear, is at what point man came to recognized the true nature of the planets, which, aside from being brighter and moving independently, were clearly very different from the stars themselves.

Yet even the oldest records acknowledged the planets as a group of related objects, with the primary issue not so much themselves, but whether they revolved around the sun or the earth; and by implication, rather than stated, that they were also 'solid bodies' like the moon.

Early heliocentric models, however, were commoner than is usually accepted, for example by Aristarchus as early as 270 BC, while the Indian astronomer, Aryabhata, in 499 was the first to included elliptical orbits in a heliocentric model, and also the first to note that 'bright' planets did not shine by themselves, but by reflected sunlight.

Ptolemy's star catalogue, in 144 AD, entrenched the geocentric universe for the next 1500 years, but it was publication of Copernicus's magnum opus, 'On the revolutions of the heavenly bodies' in 1543, which finally settled the argument, and restored the heliocentric solar system, which has persisted to this day, with the earth retaining it's privileged position as central to, but not the center of, the universe.

5.2 Origin and Formation:

The current theory to account for the origin of the solar system is the nebular hypothesis, in which the sun and planets were formed about 4.6 billion years ago from the gravitational collapse of a gigantic molecular cloud of dust and gas - mainly hydrogen and helium,

This was gravitationally unstable, and fragmented into large discrete masses, destined eventually to become stars, and in the process acquiring rotating discs of dust and debris (proto-planetary discs) up to 2,000 AU across, which would eventually give rise to planets. These were essentially accretion disc, where hot denser matter from the inner regions would spiral inward under the gravitational influence of the central mass, while the cooler outer areas would remain largely gaseous.

Particulate debris from the inner region would then clump together and build up, to form planetary embryos. This would probably happen relatively quickly initially, but it would take many hundreds of millions of years thereafter to eventually form the four ('rocky') inner planets.

Formation of the outer planets ('gas giants'), is not well understood, though it is essentially a two stage process. A rocky central core had to build up first, and begin to accrete gaseous material from the surrounding disc, and this would then eventually form the greater part of the planet's bulk.

This can only begin, however, at a sufficient distance from the sun (about 5 AU - the ice line) to be cool enough for icy planetary embryos to form, which can then grow and enlarge to many times the size of the earth. Thereafter, gaseous accretion proceeds extremely slowly at first (several million years) to acquire only a small part of a planet's bulk, and then relatively rapidly to accumulate the remainder of it's atmosphere, over as little as 10,000 years.

At this stage, however, planets have not yet necessarily settled into their definitive orbits, and are still free to migrate, sometimes over quite large distances, while these changes are taking place.

The solar system probably took about 50 million years to form, by which time temperatures within the core of the proto-sun were sufficient to initiate thermonuclear fusion, and once the opposing pressures of radiation and gravitation had equalized, to established stability, the sun then became a main sequence star.

In addition to electromagnetic radiation the sun also radiates a continuous stream of charged particles (solar wind) which enveloped the whole planetary system in a tenuous cloud (heliosphere), sweeping away the remaining dust and gas, and bringing planet formation to an end.

The sun and the solar system will then remain much as they are for about 4½billion years, until the sun exhausts it's supply of hydrogen. It will then expand slowly, doubling its sizes over the next half billion years, to become a subgiant, and then more rapidly to become a red giant, 200 times its present size, over a further period of about half a billion years.

5.3 Cosmography (Our place in the universe):

The term derives from 'cartography', and covers general descriptive features of the large scale universe, from galaxies to planetary systems[1].

The sun's position within the universe at large, which has no boundary, is not something we can define. There is no 'coordinate grid', and nor is their any central point, for space is homogeneous an isotropic throughout, and would look the same in all directions from any chosen point.

These facts are enshrined in the Copernican principle[1], which states that the Earth has no central position within the universe, while a more general version, but expressing the same sentiments, states that human beings (and by implication, any intelligent life forms) have neither favored position or status within the universe.

We could, if we choose, simply allocate a position for the sun, with reference to some nearby galaxies, but there would be nothing permanent about it, in an environment in which all galaxies are receding ever more rapidly from one another.

The situation is different within our own local galaxy, however, (the Milky Way) to which the Solar System belongs, and where we can easily specify the sun's position with respect to any number of internal markers; and indeed the sun's exact position within the Milky Way has important implications for the presence and evolution of life on earth, and we discuss these aspects further in chapter 7.

The Milky Way is a barred spiral galaxy, with a diameter of about 120,000 light-years, and estimated to contain around 200 billion stars, concentrated into 4 main spiral arm. The cental region (galactic bar) is consistent with the presence of a super-massive black hole, about 4 million times the mass of the sun. The sun and solar system lie about 25,000 light years from the galactic center (Sagittarius A) on the rim of the Orion arm (galactic habitable zone). They rotate with the galaxy, and complete one revolution approximately every 230 million years (figure 5.1).

For most of this cycle, the solar system lies largely outside areas of high stellar density (where supernovae abound, together with gravitational instability and high levels of radiation) and also,

Figure 5.1

Position of the solar system (!) within the Milky Way galaxy, in relation to the central bar, and two of the main spiral arms . It's period of rotation (1 galactic year) = 230 million earth years.

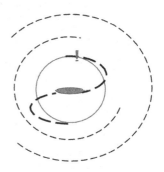

because the sun orbits the galactic center at roughly the same speed as the spiral arms, it only passes through these regions infrequently.

The plain of the ecliptic (in which the planetary orbits lie) is inclined at 600 with respect to the plane of the galaxy.

5.4 Structure and Content[2] (of the Universe, 2.6):

The Sun (5.7), a G2 main sequence star, is the central feature of the solar system. It makes up over 99% of the total mass, while the four largest planets (Jupiter, Saturn, Uranus and Neptune - known as the 'gas giants') make up 99% of the remaining mass, i.e. only 0.4% of the total mass.

All the major planets have virtually circular orbits (5.5) which lie within the plane of a flat disc, known as the ecliptic.

A planetary System may be defined as a set of (non-stellar) gravitationally bound objects in orbit around a central star. Usually this will include one or more planets, together with some, or all of the following: dwarf planets, planetesimals, asteroids, natural satellites, comets and meteoroids,

The Solar System is the planetary system to which Earth belongs, together with most of the other objects listed, while more recently, some would like to see artificial satellites and space probes also included in this definition.

'Definitions' in this day and age are becoming quite problematic, with so much information available that the balance between 'size' and 'content' is not always easy. The International Astronomical Union, where many of the objects they have to deal with are themselves changing over time, have been wrestling with this problem for years, and have now decided to compile what amounts to an index of definitions, under the title:

'Three-way Categorization of Bodies Orbiting the Sun'.

Balancing 'accuracy' with 'completeness', however, can easily compromise clarity, and some of the definitions they have come up with are less than elegant hybrids. A 'Planet', for example is:

A celestial body that (a) is in orbit around the Sun, (b) has sufficient mass for its self-gravity to overcome rigid body forces so that it assumes a hydrostatic equilibrium (nearly round shape), and (c) has cleared the neighborhood around its orbit[3].

The Solar System consists of the Sun and eight major planets: four inner ('terrestrial') planets, with rocky cores - Mercury, Venus, Earth and Mars; and four outer planets (gas giants) - Jupiter and Saturn, primarily composed of Hydrogen and Helium, and Uranus and Neptune (the ice giants) composed of water, ammonia and methane.

Large numbers of 'small' to 'intermediate' sized objects are also present, and tend to be confined to belts within the solar system, e.g. asteroid belt, between Mars and Jupiter, and the Kupier belt, just beyond

Neptune. Previously, these have variously been described as planetoids, minor planets or asteroids, but most have now been reclassified as dwarf planets - defined as objects which orbits the sun independently, are not gravitationally related to other objects, and their shape is determined solely by their own gravitation. This definition distinguishes them from natural satellites (many of which have similar physical characteristics) which circle planets in orbits which are gravitationally determined by the mass of the primary. Currently (December 2013) there are 422 natural satellites,173 of planets and 249 of minor planets.

Uranus, discovered accidentally by William Herschel in 1781, was the first major addition to the historical solar system of naked eye planets, and subsequently Neptune was the first object to be discovered based solely on theoretical prediction.

The solar system today is now very different from the past. Many new 'objects' and 'regions' have been discovered and added, some ill defined and overlapping, and with considerable confusion in terminology, as definitions change and objects are reclassified.

The traditional division between 'inner' and 'outer' regions, however, based on the physical characteristics of the major planets, has been retained, Figure 5.2.

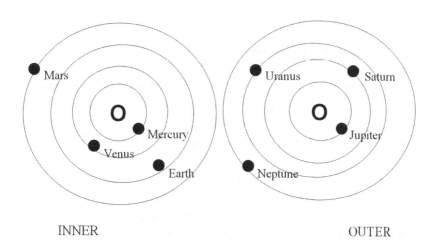

INNER   OUTER

Figure 5.2: Main divisions within the Solar System:
Inner (terrestrial planets): Mercury, Venus, Earth, Mars, Asteroid belt.
Outer (gas giants): Jupiter, Saturn, Uranus, Neptune and Kupier belt.

The Asteroid belt is the most familiar collection of 'smaller objects', ranging in size from 'minuscule', up to the size of Ceres (the largest) which is about 600 miles in diameter. It lies between Mars and Jupiter, 2.3 - 3.3 AU from the sun. Asteroid objects are now known as 'small solar system' objects, with the exception of Ceres itself, which was reclassified as a 'dwarf planet' in 2006.

The trans-Neptunian region marks the outer extremes of the solar system. It is relatively more rarefied, but still contains large numbers of smaller objects. The Kupier belt (between 30-50 AU from the sun) is a fairly well defined torus of small ices, including the dwarf planet Pluto, while the scattered disc, which overlaps it, is thought to be the source of many short period comets, with some having their aphelia as far out as 150 AU.

The heliosphere of charged particles (solar wind) permeate throughout the solar system, and terminates at the heliopause, between 100-200 AU out, which marks the boundary between the solar system and interstellar space.

The Oort cloud is a hypothetical spherical region extending outward for up to 1½ LY, and the likely source of long period comets, such as Halley's, while, by comparison, the sun's actual gravitational influence can probably extend as far out as 2 LYs (125,000 AUs).

5.5 Orbits (Bode's Law):

Throughout most of antiquity, planetary movements per se were of little interest, compared with the association between the planets themselves and the Gods, while the popular image of being attached to the inside of a revolving crystal sphere persisted well into the renaissance. Nevertheless, 'orbits' are mentioned in records dating back a far as the 2nd millennium BC, when even the nature of their geometry was being questioned.

Prior to Copernicus, however, concerns were mainly with whether planets circled the earth or the sun, than any of their physical characteristics, but thereafter, the heliocentric model, though cumbersome initially, gradually became accepted.

Kepler was the first observer to appreciate the importance of 'geometry', though only after two abortive years devoted obsessionally to studying circles (as one of nature's 'perfect forms'), when he reluctantly turned his attention to 'ellipses'. We know today that all orbits are elliptical, and indeed it is a mathematical necessity that they should be so (5.6).

For descriptive purposes, however, the orbits of all of the major planets can be regarded as circular. Pluto alone is moderately eccentric, while the orbit of Sedna (a dwarf planet, and very recent addition to the solar system) is a grossly elongated ellipse, measuring some 500 x 170 AU. Comets, without exception, have elliptical orbits, often extreme, and these can be attributed to the nature of their formation out of dust and debris, and their low density susceptibility to gravitational influences. Comets are now classified as 'small solar system bodies'.

Remembering 'distances', however, is not easy, and in the 17th century, the astronomer Johann Bode discovered a very useful aide memoir. This was an empirical relationship, in the form of a numerical series which closely approximates to the distances between the Sun and each of the 6 known planets at the time, but it was actually one of his contemporaries who gave the first clear account of how it was derived.

He visualized the radius of Earth's orbit as divided into 10 equal units, and on that basis, the distance of Mercury from the Sun would be 4 units; Venus, 4+3 units; Earth 4+6 units, etc....extending by doubling the previous number of units each time, to give the series:

i.e.

4	4+3	4+6	4+12	4+24	4+48	4+96
4	7	10	16	28	52	100

Dividing each of these numbers by 10, and the figures then closely approximate to the distance of each planet from the sun (Bode's Law):

0 .4   0.7   1.0   1.6   2.8   5.2   10.0   19.6   39.2

Planet:	Me	V	E	M r		J	S	U	N
True distance from Sun	0.39	0.72	1.0	1.52	(2.77)	5.20	9.54	19.2	30.0

Alternatively, this relationship can be expressed more concisely by a simple formula:

$$d = (0.4 + 0.3 \times 2^n)$$

where d is distance from sun, and n = 0, 1, 2, 3.......

Bode's law was accurate to within 1-2%, for 5 of 6 known planets, but particularly came to prominence with the discovery of Uranus in 1781, which fitted the pattern so closely, that it proved an incentive to search for a 'missing planet' (2.77) between Mars and Jupiter, and the eventual discovery of Ceres in 1801.

In spite of theoretical proposals (gravitational interactions, for example) no explanation for Bode's relationship has ever been found, and it is now regarded as little more than a curio, having fallen out of favor with the discovery of Neptune (with a predicted value almost 30% too high) and subsequently the discovery of asteroids, minor planets, the kuiper belt, etc.

5.6 Kepler's Laws of Planetary Motion (1604)[4]:

Tycho Brahe was the last of the great naked-eye astronomers, and his meticulous work over the years left records of unsurpassed accuracy. Kepler was initially Tycho's assistant, and on his death, rather than continuing to collect observations, he took over the task of trying to analyze and explain Tycho's results in terms of geometry. This was the beginning of 'precision astronomy', and the data on which Newton would later found his celestial mechanics.

All previous models of the solar system, whether geocentric or heliocentric, were nevertheless based entirely on circles - complex combinations of large circles, with smaller circles (epicycle) revolving around their circumference, and in theory, with a suitable combination, any pattern of planetary movements could eventually be explained. Indeed, even after Copernicus had finally established heliocentricity, the actual model he employed was more complex, and less accurate, than the contemporary geocentric model it was replacing.

Nevertheless, the old system had been popular for very good reasons - every major culture, since earliest historical times, believed the earth to be the center of the universe, and it would be inconceivable if this was not also true for planets.

The new Copernican alternative assumed three things - that the planets revolved around the sun in circular orbit; that the sun was at the center these, and that the planets moved with uniform velocity

Kepler had reservations however, and after almost 2 years, working solely by 'trial and error', he was forced to abandon circles, and turned his attention to ellipses. At about that time he also completed a particularly careful study of the movements of Mars, and it was soon clear that an ellipse was much more likely to fit his data than circles had ever been, Figure 5.3.

An ellipse is the closed curve formed by the intersection of an inclined plane with the walls of a cone. The foci are two fixed points (F1, F2) on the major axis of the ellipse that are equidistant from the center point, and the sum of the distances from any point on the circumference to these two foci is constant, and equal to the major axis. The sun lies at one focus, and the center of the ellipse is the

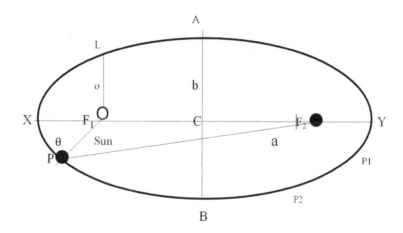

Figure 5.3: Geometry of an ellipse (orbital period P): XY-Major Axis(2a); AB-Minor Axis(2b); F1 F2 - Foci; F1 L - Semi Latus(ø). Eccentricity is the ratio of the distance between the two foci and the major axis of the ellipse, F1 F2 / XY; PF1 + P F2 = XY = constant.

midpoint of the line between the two foci.

Owing to it's oblique shape, the distance between the sun and any planet moving on the circumference is constantly changing, with the point of nearest approach to the sun known as perihelion (X), and the corresponding point, at the opposite end of the curve, where it is furthest from the sun, the Aphelion (Y); and as a corollary to this, the orbital velocity of the planet cannot be uniform, as the Copernican proposals envisaged.

In Figure 5.3, the angle X F1 P (θ) is the planet's present position, measured anti-clockwise from the perihelion.

a, b are the semi-major and semi-minor axes, respectively,

ø the semi-latus (F1L), and e the eccentricity.

Kepler established his three laws of planetary motion by matching Tycho Brahe's observational data with the properties of ellipses rather than circles, and his first law was essentially a statement of confirmation:

First Law: Planets move in elliptical orbits, with the sun at one focus.

Nevertheless, the orbits of all major naked-eye planets are so close to circular, that this fine distinction was part of Kepler's initial difficulty, especially where empirical 'matching up' was concerned.

Second Law states that the line joining a planet to the sun, sweeps out equal areas in equal intervals of time.

Third Law states that the square of the orbital period is proportional to the cube of the semi major axis (Fig 5.3)

$$P^2 \propto a^3$$

The validity of Kepler's laws, and the third law in particular, was questioned when they were first proposed, in part because there was no apparent justification as to why they should hold true, and it would be 70 years before Issac Newton published his theory of gravitation, which fully explained Kepler's relationships, and laid the foundations of celestial mechanics.

5.7 The Sun:

The sun is a second generation star, which explains why it arose relatively late-on in cosmic history - 4.6 billion years ago (9 billion years after the Big Bang). It formed initially from gravitational collapse within a massive molecular cloud (4.8). With continued contraction and compression, the core temperature rose steadily, until it was hot enough to initiate thermonuclear fusion, converting hydrogen to helium, and this is the source of power and energy within all stars for the greater part of their lifetimes[2].

The sun is composed almost entirely of hydrogen (74.9%) and Helium (23.8%), while metals collectively make up less than 2%, primarily oxygen, with traces of carbon, neon and iron. It converts 620 million tons of Hydrogen into Helium every second, and so long as it continues to derive it's energy in this way, it is classed as a G type main sequence star (which specifies its spectral type and luminosity).

Large massive stars are relatively uncommon (the majority in our galaxy are red dwarfs), but the sun nevertheless is larger than the average star, and brighter than 85% of stars in the Milky Way. Also, it has still only consumed about half it's original mass, and so has a further 4½ billion years to go before it's hydrogen runs out, when it will expand to become a red giant; in the very long term it will end up as a white dwarf [5].

Physical characteristics[2]:

Parameter:	Sun:	Earth:	x Earth:
Diameter	870,430 mi	8,000mi	109x
Mass	9.8 x 1027Tons	5.8 x1021Tons	333,000x
Surface Temp.	5,778 K	288 K	20x
Surface Gravity	28 g	1 g	28 x
Escape Velocity	385 mps	7mps	55x

Structurally, the sun is an enormous spherical ball of hot ionized gasses (plasma), interspersed with intense magnetic fields, and will retain it's spherical equilibrium for as long as the outward pressures of radiation exactly balance the inward forces of gravitation. It's gaseous nature also accounts for continuous internal convection activity, which both distributes energy and explains the sun's differential rotation rates - faster at the equator than the poles, but with an average period of about 28 days.

The sun has a layered internal structure. The central core extends outward for about 25% of the radius, with a temperature of many millions of degrees, and is the only region where fusion reactions take place and energy is actually generated. This is surrounded by a narrow radiation zone, where ionized plasmas transfer this energy outward. The remaining 70% of the sun's radius, forming the bulk of it's mass, is the convective zone, where lower temperatures mean that gases are no longer fully ionized and radiative heat transfer is therefore reduced. However, because gases are now correspondingly less dense, this compensates by allowing extensive convective activity throughout the whole of the sun's bulk, and transfers thermal energy outwards and upwards to the sun's surface.

The photosphere is the visible surface of the sun, which is no longer opaque, and allows light and energy to propagate into space. It is many thousand of miles thick, but much less dense than the corresponding regions of the earth's surface, and has a spectrum which closely approximates to that of a black body at a temperature of about 6,000 K.

Anything beyond the photosphere is known as the sun's atmosphere, and includes the chromosphere, corona, and heliosphere, extending out as far as the orbit of Pluto and the heliopause (5.4), which defines the transition between the sun's domain and interstellar 'space'.

The sun is also a source of powerful magnetic fields, permeating throughout it's whole structure, constantly changing, and reversing polarity every 11 years. This activity in turn is reflected in a wide variety of visible solar activity - sun spots, flares, prominences, the solar wind, and indirectly, aurorae and electrical storms on earth.

Sunspots are among the most impressive aspects of solar activity, typically with a black central region (umbra) surrounded by dark irregular halos (penumbra); these correspond to sources of intense magnetic activity, but the dark regions are far from cold, and simply reflect temperature differentials. Sunspots tend to distribute along parallel belts of latitude, which shift but never encroach onto the polar regions. Their numbers vary cyclically over the 11 year period, from sometimes none at all, at 'minimum', to hundreds covering much of the sun's surface, at 'maximum'.

Meteorologists have long sought to correlate sun spot activity with terrestrial weather, with some suggestive coincidences, but nothing has ever been proved.

5.8 Planets:

(1) Standard Features:

	Orbit (Radius AU)	Diameter (Miles)	Mass (x Earth)	Period Sidereal / Orbital	Surface temp C	Escape velocity	Satellites
Me	0.4	2,780	0.05	58d / 87d	67	2.2 mps	0
V	0.7	6,898	0.8	-243d / 225d	460	5.8 mps	0
E	93million miles	7,262	588x1024 to ns	24h / 365d	15	6.2 mps	1
Mr	1.5	3,848	0.1	24½h / 686d	-63	2.8 mps	2
J	5.2	81,334	318	10h /11yr	-108	34 mps	67
S	9.5	68,262	95	10½h / 29yr	-140	20 mps	62
U	19.2	14,520	14	17¼h / 84yr	-200	12 mps	27
N	30	28,230	17	16h /165yr	-200	13 mps	14

(2) Special Features:

Mercury: Smallest planet; Negligible atmosphere; Heavily cratered surface, similar to moon, from very active volcanic past.

Venus: Retrograde rotation; CO_2 atmosphere, 90x denser than earth; Hottest planet (greenhouse effect); active volcanic activity.

Earth: Dense oxygen atmosphere (21%) - a long term consequence of life (7.8); geologically active; only planet with surface water and plate tectonics (continental drift); only known source of life;

Mars: Very tenuous atmosphere, mainly CO_2; covered with volcanoes and rift valleys, but no longer geologically active; Frequent dust storms; red color derived from iron oxide (rust) in the soil; present surface exploration shows no evidence of life.

Jupiter: Largest planet; more than 2x the total mass of all other planets; composed of H and He, with no solid surface, though possible a small dense central core; equatorial regions rotate at more than 28,000 mph, making Jupiter's disc an oblate spheroid; temperature rises inwards with compression, to over 30,000 K at the core, but Jupiter would need to be at least 75x more massive to progress to fusion and become a star; the outer surface is covered with thick cloud layers, in parallel bands of latitude, manly composed of ammonia crystal, and embracing all levels of atmospheric activity, from lightning, to areas of extreme turbulence, such as the great red spot - a semi-permanent region of anticyclone storms, several times the diameter of the earth in size; Jupiter has the strongest magnetosphere of all the planets, and also a very faint ring system (discovered by Voyager 1, in 1979).

Saturn: Has many similarities to Jupiter, but at only 60% of the volume, is the least dense planet in the solar system; the rings (composed of 93% ice crystals, and the rest tholins†) extend outwards from 3,800 to 68,000 miles, but are barley 18ft thick.

Uranus is the coldest of the planets, (- 224^0 C), with a complex layer structure, enclosing a small central core of ices and rock, and possibly some water. It's rotation axis is tilted by over 900.

Neptune is 17 times larger than earth. The densest of the four gas giants, it is slightly warmer and smaller than Uranus, but otherwise very similar in composition.

5.9 Small Bodies and Outer Regions (5.4):
Collectively, these features are a minefield of definitions and overlap. All objects other than planets, satellite or dwarf planets are now classified as 'small bodies'.

Dwarf planets include Pluto and Ceres, and about 200 other objects, mainly in the asteroid belt[6].

All comets are now classified as 'small bodies', with short period comets thought to originate in the Kupier belt and scattered disc, and long period comets in the Oort Cloud.

---

† Molecules formed by UV radiation of organic compounds such as methane.

Detached objects are a separate trans-Neptunian group, distinguished by extremely elongated orbits, such that they are not subject to the gravitational influences of Neptune. Sedna is the best know of these, and the largest trans-Neptunian object. It has the longest period (11,400 years) and most distant aphelion of any object in the solar system - at 937 AU, sometimes taken to indicate the size of the solar system.

The heliopause, is the outer limit of the sun's influence (heliosphere), and marks the transition where the solar wind tails off into interstellar space. Both Voyager 1 and Voyager 2 are reported to have reached the outermost regions of the solar system, and Voyager 1 reached the heliopause and passed into interstellar space in December 2012.

5.10 Habitable Zones[7]:

We look at the nature and properties of life in Part 11, but paramount among these is not only the wide variety of conditions required (chapter 7), but also the way in which these relate to and interact with each other, for nowhere is homeostasis more important than with respect to a life-supporting environment.

The extreme range of physical conditions within the solar system represents the archetypal properties of any circumstellar environment, and especially with respect to the central star itself, which exerts an influence throughout the whole of the interplanetary domain, and thus to a large extent determines it's eventual suitability as a habitat for life.

The Habitable Zone (7.5) is defined as that region around a star within which planet-sized objects have an atmosphere capable of supporting liquid surface water; indeed, so importance is water to supporting life, that the existence of such a region is now one of the major criteria for the presence of extraterrestrial life (chapter 15).

The concept of a circumstellar habitable zone (CHZ) developed in the early 50s, together with related terms such as 'ecosphere' and 'Liquid Water Belt'. Based on these 'new criteria', an early estimate put the number of potentially habitable planets in the Milky Way at about 600 million, but by 2013, based on results from the Kepler space mission, this figure had risen to 40 billion earth-sized planets orbiting sun-like stars, with the nearest of these barely 12 light years away.

On a much larger scale, an analogous region has been defined with respect to the galaxy, where the galactic habitable zone (5.3) is that region of a galaxy where life would be most likely to evolve - close enough to the center for stars to provide the necessary heavier elements (i.e. other than H and He), but not so close as to suffer the consequences of radiation and of excessive gravitation.

Many factors determine whether a planet will be in the habitable zone of it's parent star, but common to all are radius of its orbit, mass (size) of the planet, and the prevailing levels of radiation. Others include magnetic fields and atmospheric characteristics, such as 'greenhouse effects', and overall two different habitable regions are now distinguished, extending outward to 1½ AU and 3AU respectively from the sun.

A number of planets and larger satellite also have habitable zones, and the same general principles apply to evaluating these.

The habitable zone of a star is related to it's luminosity, and therefore can change over the lifetime of the star, and to allow for this, the term 'continuously habitable zone' has been introduced, as the region surrounding a star which can sustain liquid water for a given period of time only.

However, for main sequence stars, such as our sun, once it is established on the main sequence, it's luminosity will not change significant over many billions of years. Hence, it may well be, that if life is destined to develop, it will nor do so until a star enters a long period of evolutionary stability, though we should also bear in mind that once life has 'taken root', it evolves so gradually thereafter, that even if the environment conditions change significantly, it may still be able to adapt.

If such conditions appear to be restrictive, they are no more so than some of the 'preconditions' which seem to apply to 'fine tunning'(7.3).

An inevitable feature of habitable zones is water depletion over time, and an absolute requirement for any 'habitable environment' is that there must be some means available to replace water loss; we can hypothesizes about many potential mechanisms where earth is concerned - impact with ice bodies from outer space, for example - but none so far are adequate, and the reality is, that we still do not know where the earth got it's water from in the first place[8] (7.6).

The search for extra terrestrial planets is now continuous 24 hours a day, with automated equipment scanning stars for tiny changes in brightness, which might indicate an object in transit across their disc. We discuss the detection of exoplanets in chapter 12.10.

5.11 Exploring the Solar System:

The first man-made object to enter space was a V2 rocket in October 1942 - the culmination of Germany's V2 project at the end of WW2, and the beginnings of rocketry and space technology.

After the war, both Russia and America began to develop active space programs, while Wernher von Braun, and most of his team of rocket scientists, eventually took up residency in the United States, and were responsible for much of the pioneering work on rocket propulsion.

Early Years:

Almost from the outset, America's space projects have been under the auspices NASA, established by president Eisenhower in 1958, largely to encourage peaceful applications of space science, and shifting the emphasis from military to a mainly civilian based organization.

Early projects included the Apollo missions (6 manned lunar landings between 1969-72); Skylab (1973/74); Space Shuttle (135 missions between 1981 and 2011), and currently the International space station.

Russian space programs were severely compromised for years, due to political instability throughout the USSR, with many early projects reflecting the need to keep abreast of the US, and a desire for 'prestige'. These included Sputnik (1957, first artificial satellite); Yuri Gagarin (1961, the first man in space); Soyuz (orbit shuttles, from 1967 onwards); the Mir space station (1986-96) and the Buran Space Shuttle (1986-96) - a 'copycat' version of the American shuttle, it made only 1 (unmanned) flight over it's entire lifetime.

The Federal Space Agency (equivalent to NASA) was only formed in 1992, and currently the Russians are active participants in the International Space Station.

Planetary Explorations:

Many interplanetary flybys were undertaken between 1962 and 1989, including at least one mission to every major planet in the solar system, and most have now been visited a number of time since, together with some of their satellites.

Mercury is the least well explored of the major planets, The MESSANGER space probe went into orbit around mercury in 2011, after a journey of over 6 years, with many complex manoeuvers, including 'gravity assist' to get there.

The first interplanetary surface mission to Venus (1970) lasted only 23 minutes, but there have been many successful flybys and soft landings since, allowing the surface to be mapped both directly and by radar; one satellite was also successfully programmed to enter and directly examine the thick dense CO2 atmosphere which Venus possesses

The Russians were the first to explore Mars, with a series of probes, between 1960 and 1970, overlapping the 9 US Mariner probes, and later the Viking program in the mid 70s. Since then, Mars has been the target for dozens of robotic space craft, orbiter, landers and rovers, from all four major 'space' nations - US, Russia, Europe and Japan, but this popularity has come at a huge cost, with two-thirds of all missions ending in failure.

The Most notable Mars missions have been two NASA landers, Opportunity, the Mars Explorer Rover, which landed in 2004, and is still active in 2014, 90x its expected useful life; and Curiosity, a robotic mobile science laboratory, about the size of a small car, which landed in August 2012, and has been active on the surface ever since, with an extensive itinerary, including the search for microbial life, climate and geology, subterranean water, and to evaluate 'habitability' for subsequent manned explorations. So successful has this mission been, that it has now been extended indefinitely.

Nevertheless, in spite of so many failures, prospective (volunteer) candidates are already undergoing training, including simulation, for a manned journey to Mars, possibly within 15 to 20 years; and most remarkable of all, there are some among them who would even be willing consider a one-way journey!

A large number flybys missions to Jupiter have been made since 1973, including recording the impact of Comet Shoemaker–Levy 9 in December 1995. Voyager 1, a space probe to explore the outer solar system, was launched in 1977, and made an extensive study of the Jovian system, including it's Galilean moons, and especially Europa and Ganymede, where future soft-landings may be attempted to explore the possibility of sub-surface water.

Although Jupiter may have a small central rocky core, gravitation would preclude any attempted landing.

Voyager 1 went on to explore Saturn in 1980, including it's ring system, and a flyby of Titan, before progressing on to the outer solar system.

Saturn was also explored by the Pioneer probe in 1979, Voyager 2 in 1981, and by the Cassini–Huygens space probe in 2004; this latter also carried out a close flyby of Phoebe, and subsequently the Huygens probe completed a soft landing on Titan in January 2005.

Titan has a thick dense atmosphere, much denser than that on earth, and which contains both methane and ammonia, very similar to the hypothetical pre-biotic atmosphere (7.9) in which life may have originated. This raises fascinating issues with respect to panspermia, and we return to these in chapter 9.

Uranus has been the targeted of only one mission, Voyager 2 in 1986. It's atmosphere was unremarkable, but it's magnetosphere was unique, and profoundly affected by the planet's unusual axial tilt (5.8). The moons of Uranus were also surveyed, and Miranda showed evidence of having undergone unusual geologically active in the past.

Voyager 2 also went on to survey Neptune, in 1989, which was found to have an extremely active turbulent atmosphere. It also discovered 5 previously unknown moons, and 900 tenuous rings, together with a number of partial rings.

There have been no space missions to Pluto, which so far remains unexplored.

In August 2012 voyager 1 crossed the heliopause (5.4) at a distance of 121 AU, and exited the solar system into interstellar space, though signals from it have still been detected even since then.

5.12 Landmark Projects[9]:

Space exploration has expanded progressively since it's inception in 1957, and the following list is a historical summary of landmark achievements over the first 55 years of human expansion into space:

1957:
Sputnik 1 - October, first satellite in earth orbit.
Sputnik 2 - November, first animal in space.
1959:
Luna 1 - January, first lunar flyby.
Luna 2 - September, first lunar impact.
1961:
Vostok 1 - April, first manned orbit of earth.

1962;
    Mariner 2 - August, first planetary encounter - Venus flyby.
    Mars 1 - November, Mars flyby.
1965:
    Venera 3 - November, planetary surface landing (Venus).
1966:
    Luna 1 - January, lunar landing.
    Luna 10 - March, lunar orbit.
1968:
    Apollo 8 - December, manned lunar orbiter.
1969
    Apollo 11 - MANNED LUNAR LANDING.
1970:
    Venera 7 - August, Venus landing.
1971:
    Mariner 9 - May, Mars orbiter.
1972:
    Pioneer 10 - March, Jupiter flyby.
1973:
    Pioneer 11 - April, Saturn flyby.
    Mariner 10 - November, Mercury flyby.
1977;
    Voyger 2 - August, Uranus/Neptune flyby.
1978:
    ISEE-3 - August, comet flyby
1989:
    Galileo - October, asteroid flyby.
1996:
    NEAR Shoemaker - February, asteroid landing.
1997:
    Cassini-Huygens - October, Saturn lander.
1999:
    Stardust - February, comet coma sample, collect and return.
2001:
    Genesis - August, solar wind sample and return.
2003:
    Hayabusa - May, asteroid sample and return.

2004:

MESSENGER - August, Mercury orbiter.

2006:

New Horizon - January, Pluto, Charon, Kupier belt.

2007:

Phoenix - August, Mars lander.

2008:

Chandrayaan - October, water discovered on moon (2009).

2010:

Solar Dynamic Observatory - February, solar monitoring.

2011:

Mars Science Laboratory - November, Curiosity rover
(SEARCH FOR LIFE)

2013:

Hisaki - September, Planetary atmosphere observatory.

Mars Orbiter Mission- November

Change 3 - December - lunar rover

References (Chapter 5):

1. T. Ferris, The Whole Shebang (Weidenfeld & Nicolson, 1997).
2. Wikipedia article (12 November 2013), Solar System.
3. Wikipedia article (22 November 2013), IAU definition of a planet.
4. P. G. Bergman, The Riddle of Gravitation (London, John Murray, 1969).
5. S. Hawking, The Universe in a Nutshell (Bantam Press, 2001).
6. Space.Com: Asteroids - Formation, Discovery and Explorations) (Access: Google - Asteroids).
7. Google: Habitable Zones..
8. Wikipedia article (12 June 2013), Origin of Water on Earth..
9. Wikipedia article (4 August 2014), Timeline of Solar System exploration.

PART 11

LIFE

(Chapters 6 - 10)

An extremely improbable event.
Uniquely diversified.
Created out of nothing?
Prototype for life elsewhere?

Chapter 6

LIFE

Life is thought to have arisen spontaneously out of non-living matter, about 3.5 billion years ago. The mechanism of this (abiogenesis) is not known, and attempts to simulate it in the laboratory have so far been unsuccessful (8.4).

6.1 The Nature of Life:

There are a number of ways in which we can think of 'life'. As a generic term, it includes all living things; alternatively, it can be thought of as a concept, rather than a 'physical entity', embracing a number of functional properties, of which the ability to preserve and promote self-existence is paramount. It is how these interact with each order, however, and with the external environment, which constitutes living,

Attempts have also been made to define life in terms of the properties it possesses, but there are such a large number of these, that this too has proved impractical; nevertheless, the ambiguous status of things such as viruses or desert varnish (14.10), purely reflect the lack of a proper definition.

Other approaches have tried to identify one single essential property, such as reproduction, but there are then too many qualifications to take into account, and this approach has not been successful either.

Most definitions now place emphasis on two main features:

(1). Physiological functions, all or most of which, a living organisms will display, including:

- Homeostasis: Regulation of the internal environment
- Organization: Structurally composed of one or more cells.
- Metabolism: Optimal handling of energy.
- Growth: Progressive Increasing in size and development.
- Adaptation: Change over time in response to the environment.
- Response to stimuli: Often expressed by movement.
- Reproduction, whether sexual or asexual.

All of these have an underlying physical and chemical basis for the functions which they subserve.

(2). Biophysical characteristics categorized under four headings, (Table 6.1). There is little 'overlap' between these groups, which include not only physical features, but psychological and abstract properties also:

Physical	Functional	Potential	Abstract
Unique and permanent Features:	Spontaneous b ehavior: (W hat we DO)	R eserve a bilities: (What we CAN do)	Subconscious qualities of Personality:
Ethnic	Consciousness	Communicate	Self-awareness
Height	Sleep	Activity	Individuality
Weight	Respiration	Feeding	Intelligence
Blood group	Metabolism	Speech	Memory
Finger prints	Movement	Bowels	Emotions
DNA	Reactio ns	Bladder	Insight

Table 6.1
Physical and Personality profile of human characteristics.

There is a high correlation between many of these and advanced life-forms, such as mammals and human beings, but we need something much more concrete for definition purposes. Most of those properties listed undoubtedly apply to ourselves, and an outsider observer would have no difficulty in knowing which species compiled this list, but for definition purposes, we have to remember that man comprises less than 1% of all known life forms, and it is trying to reduce the other 99% to a minimum number of common denominators that is where the problem really lies.

Nevertheless, having failed to reach a satisfactory compromise about ourselves, that has not stopped us speculating about extraterrestrial life, about which we know nothing at all, and even attempting to draw parallels between them and ourselves.

Once we get an answer to Fermi's question, then perhaps we can go down that road - depending on what the answer is - but in the meantime, extraterrestrials have done so much to popularize science fiction, that until we have concrete evidence to the contrary (e.g. SETI) that is where they belong.

H. G. Well probably marked a turning point in 'modern' science fiction, when he introduced time travel as a means of exploring the future (chapter 13), and this was a very skillful blend of two things we know so little about, but it did focus attention on the future itself, which according to Moore's law, is approaching ever more quickly, and with it the need to understand more and speculate less.

Science fiction has long claimed to reflect the reality of the future, and we have no reasons to dispute that; indeed, quite the reverse, for artificial intelligence is with us already, and will almost certainly play a more important role in the future; while the future role of artificial life, which is still in it's infancy, will probably be insignificant.

6.2 Classification of Life:

Classification is useful for many reasons. Mainly, however, it establishes a pattern of evolutionary continuity, and by tracing that backwards, we can correlate biological evolution with the changing geophysical environments. This also gives us an insight into the environmental conditions in which life first appeared, and a physical basis for theories of origin, under very different circumstance from those prevailing on earth today (8.3).

The oldest rocks date back to within 500 million years of the earth's independent formation, with the earliest evidence of life appearing some 300 million years after that. The fossil record of biological evolution dates from that time, and although there are still gaps, the overall pattern is consistent with all currently available scientific knowledge, and Darwin's theory of sequential evolution, survival of the fittest' and genetic inheritance, is now almost universally accepted.

There are still some, nevertheless, who believe that species were created independently, based on the teachings of the bible, and these views are still taught in some schools in America. They reflect deeply held religious beliefs, however, and as such are unlikely to change, irrespective of scientific facts.

The earliest known classification was undertaken be Aristotle, around 350 BC, initially differentiating between plants or animals, and within the latter group, between animals with and without red blood (i.e. between vertebrates and invertebrates), and then further subclassifications within each of these groups.

Although simple and inadequate by present standards, this remained the authoritative classification for many centuries.

The exploration of the American continent in the 16th and 17th centuries, revealed large numbers of hitherto unknown plants and animals, and was the incentive for a much more detailed classification, which evolved gradually over the next few decades.

The binomial classification, a forerunner of that which exists today, was introduced by Linnaeus in 1740. This allocated two Latin names to each specimen, the first to identify the genus, and the second the species itself.

The discovery of cells, and microorganisms changed the nature of classification profoundly, with multiple groups and changing nomenclature, through a system of grouping by 'kingdoms', and eventually leading to the 'three domain' system in use today.

Molecular biology and virology developed into separate disciplines along the way, though whether viruses are 'living' or not is one contentious issue which remains unresolved to this day.

From a biological stand point, there are so many variations of structural and function, that no single classification can ever be adequate, and while there are always going to be some disagreements, there are two main categories:

(1) Biological Classification[1]:

Groups arranged on the basis of shared physical characteristics, consistent with Darwin's principle of common decent:

Life:
    Domain,
        Kingdom,
            Phylum
                Class
                    Order,
                        Family
                            Genus,
                                Species.

Some of these use 'inferred evolutionary relations', rather than true morphology, while grouping by DNA, which has more recently come in, will reduce the number of groups even further.

(2) A Phylogenetic Tree[2]. This is perhaps an easier classification, which depicts information visually, and give a better general idea of how different groups are related:

The 'friendly simplicity' of this diagram, however, is deceptive, and belies it's true background, which is a simplified version of a speculatively rooted tree for rRNA genes, to illustrate the 3 major branches of life[2]

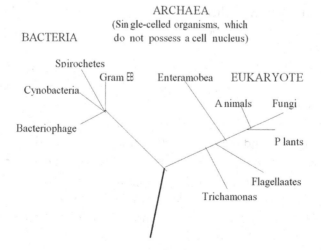

207

6.3 Concepts of Origins[1]:

[Current Theories of Origin - Chapter 8.5]

The nature of life, and what it is and means, has been a on-going topic of debate ever since Aristotle first introduced classification, which became a major topic in the 17th century, with pioneers such as Pasteur and Darwin, who laid down the fundamentals for Oparin, Miller and Urey in the 19th century.

These latter were the first to approach the question of origin as a scientific project, and did a great deal of original work in the 50 and 60, which is still widely respected even today. Indeed, there have been no further serious attempt to try to produce life under laboratory conditions, and although up to now, earlier work was largely seen as discredited, some recently published findings in 2014 now appear to supports much of Oparin and Miller's early proposals, and has brought the whole question of abiogeneses (8.5) back into the picture.

There has been no shortage of proposals over the years, however, to try to account for life, and although some of these barely seem credible in the light of present knowledge, the reality is that it is not actually possible to disprove any of them..

It is the nature of science, nevertheless, that it will always seek to explain anything which has physical existence, in terms of existing scientific knowledge, and although there are many instances where this has not yet happened, it is an implicit part of scientific expectation, that it will nevertheless succeed eventually - if not today, then whenever knowledge catches up.

In reality, however, there are only three ways in which life can be explained - scientific, outside intelligence, or pure chance, and we discuss these options later.

Religion has been the major contender over the years, but concurrently with that, four other proposals gained considerable support and prominence in their day:

Materialism is one of the oldest beliefs of all, dating from as early as 400 years BC, and expounded the principle that all things could be rationalized in terms of the four 'eternal element' - earth, air, fire and water. 'Change' was explained by arranging and rearranging these four elements, though it was never clear 'how' or 'by whom' this would be done, and with only 24 possible permutations of four items, it hardly seemed a very versatile hypothesis.

The idea of a 'soul' was something which came much later - initially in the form of 'fiery atoms'; but it was seen as the primary 'instigator' of life, and as such a concept which would persist. The 'eternal elements', however, were gradually phased out, and eventually replaced by 'parts' (unspecified), which could be assembled into animals and human, and would then function together as a machine.

Such theories, in one form or another, persisted almost up to the 19th century, when the whole concept of life changed, with Darwin's theory of progressive evolution by natural selection.

Hylomorphism was another early idea introduced by Aristotle, and was even more nebulous than 'materialism' - that everything in the universe was composed from combinations of 'matter' and 'form'. The latter was synonymous with 'soul', of which there were three varieties, each associated with different functions - vegetative (for plants), the animal soul, and the 'rational' soul, which subserved the higher functions only found in man.

However, the whole concept of a 'soul' was confusing, and certainly in conflict with the beliefs of religion, and concepts of an 'after life'.

Vitalism was slightly more scientific in it's formulation, and persisted into to 19th century on the basis of support from some of the most prominent figures of these times. It was the first proposal which saw organic matter as unique to living entities, and as such essentially different from inorganic matter. The key role of carbon, however, as the truly fundamental basis of life, was not appreciated until much later, but nevertheless, vitalism reflected a mote rational approach to understanding the chemistry of life, and was part of the transition from Renascence to modern times.

In the 1920s, Alexander Oparin became interested in the composition of the earth's early atmosphere, and proposed the first 'testable' theory, based on the concept of a 'prebiotic soup', and this led directly to the work of Miller and Urey a few decades later (8.3).

Panspermia proposes that life did not arise on earth, but rather elsewhere in the universe, and was 'seeded' into the earth's atmosphere, at a some later point in time. We discuss this hypothesis in chapter 9.

6.4 The Evolution of Life[3]:

[Ma = million years ago]

Time from today	Essential features:
4,600 Ma	Earth forms from proto-planetary accretion disc surrounding the sun; complex organic molecules may already be present.
4,500 Ma	Moon formed, following a collision with a large 'asteroid' sized object. Lunar gravitation 'stabilizes' earth's axis.
4,000 Ma	Earth's oldest rocks form in Canada
4,100-3,800	Heavy meteorite bombardment of inner planets. Thermal environment may have been conducive to early life.
3,900-2,500	Early chemotrophs oxidize inorganic materials to obtain energy. Later, energy freed e.g from carbohydrates, forms the chemical bonds of ATP.
3,700 Ma	Earliest evidence of life on earth - graphite found in Greenland, and microbe fossils in Australia.
3,000 Ma	First land life appears; atmospheric Oxygen toxic to anaerobic organisms, leading to:
2,400 Ma	Oxygen catastrophe - earth's anaerobes wiped out; one of the worst extinction event of all times.
2,000 Ma	Proliferation of single-cell organisms.
1,500 Ma	Bacteriophages. Oxygen atmosphere permanently established.
1,200 Ma	Sexual reproduction in fossil records. Multi-cellular algae.
1,000 Ma	Multicellular life.
600 Ma	Simple animals.
500 Ma	Fish and early-amphibians.
475 Ma	Land plants.
400 Ma	Insects and seeds.

360 Ma	Amphibians.
300 Ma	Reptiles.
200 Ma	Mammals.
60 Ma	Primates.
20 Ma	Hominids (great apes).
2-3 Ma	Genus Homo (human predecessors).
200,000	Modern humans.

6.5 Extinction Events:

Throughout the course of evolution, periodic extinctions temporarily reduced the diversity of life forms, eliminating:

2,400 Ma	Oxygen catastrophe - Anaerobic organism.
252 Ma	Permian-Triassic extinction - trilobites.
66 Ma	Cretaceous–Paleocene extinction - dinosaurs.

6.6 Human Evolution[4]:

The concept of life can also be thought of as generic, in the more physical sense of embracing a large number of possible 'life forms', each with it's own properties and characteristics, although sequentially linked by the chronology of evolution, such that the youngest is always the most advanced - i.e. where human life stands today.

Man's predecessors first appeared between 2-3 million years ago; in his present form, man has been around for barely a thousandth of that time, and with evolution trending toward exponential, he can expect to cover as much ground again (in terms of quality) in a fraction of that time.

The earliest changes were evidence of the use of stone tools in the fossil records from these eras, suggesting the beginnings of inductive behavior, together with bipedalism, which evolved around the same time, as an adaptation for terrestrial living.

The next million years was characterized by encephalization, when cranial capacity doubled, and between 1.3 and 1.8 million years, Homo erectus became the first homonid to expand beyond Africa, into Europe and Asia.

Over the next 500,000 years, homo erectus spread widely throughout Eurasia, and over the same period gradually evolved to become Neanderthal man.

The classification of homo sapiens, and the earliest fossils of modern humans date from that time - the Paleolithic period, about 200,000 years ago. Anatomical characteristics evolved first, and then over the next 50,000 years, gradually blended to display full behavioral characteristics. Thereafter, homo sapiens began to expand and colonize other continents - Australia by around 40,000 years ago; the Americas by 15,000 years ago, and eventually arriving 'back' at New Zealand between AD 300 and 1280.

Basic needs for survival and subsistence were ongoing priorities, with agriculture, and the 'domestication' of plants and animals, dating from about 12,000 years ago, and progressing to the beginnings of civilization - unification and expansion, development of communal life, culture and religion; formation of governments, and an eventual territorial hierarchy of states, countries and empires.

Evolution today embraces every form of progress, and we can date modern society from the rapid changes in almost every field of human activity over 19th and 20th centuries. As machines and technology came to supplement man-power, needs for power and energy escalated, and a fuel-based economy became ever more dependent on (forseably dwindling) natural resources.

Changes in science and medicine altered the whole nature of man's existence, from tolerant acceptance of a status quo, to actively maximizing current potentials, to artificially accelerate his own advancement.

Yet man could equally well be the harbinger of his own demise, for so long as the potentials for self destruction exist, the futures can never be taken for granted. By contrast, however, there are equally uncertain natural possibilities. Stephen Hawking, for example (among others), have noted that statistically the next large scale extinction event (6.5) is now overdue, in the context of a 'window' some thousands if not millions of years wide, but nonetheless inevitable eventually - the only genuinely unknown, is precisely what form it will take.

6.7 Artificial Life:

As we saw earlier, there is no precise definition of life, but rather a catalogue of properties, most (but not all) of which an entity will possess to qualify as 'living'. However, we have still not been able to specify which of these are essential, and it has been suggested that there are 'different levels' of life - an ameba, for example, can metabolize and reproduce, but little else; while man possesses all of these functions - yet both are 'life', in the fullest sense, while in all other respects they are not remotely similar.

No matter how one looks at it, life is a 'function' rather than a 'thing', and hence any true definition should concentrate on 'what it can do' and 'how it performs', rather than a catalogue of physical properties. Nevertheless, introducing new criteria alters the emphasis entirely, and perhaps simplicity at the expense of accuracy is a better compromise, so long as a definition is suitably specific to its subject. 'Life', for example, might be defined simply as:

The ability to preserve and promote self-existence

Defining artificial life, however, is even more problematic, because it depends on what 'property' or 'ability' the concept of 'artificiality' applies to. By contrast, artificial intelligence is straight forward, with no ambiguity of function, and in practical terms is more likely to be the reality of the future than artificial life itself.

Nevertheless, however we choose to define 'life', once all possible functions have been listed, one is still left with the feeling that something has been left out - an elusive 'something else', that we can neither specify or identify, but would always know if it was absent. An infallible test of artificiality, which although subjective, might even be as reliable as the Turing test.

The popular image of artificial life is some sort of humanoid robot, with general characteristics of 'man', exemplified, for example, by Mr Data of Star Trek. That would be entirely misguided nonetheless, because, as we saw earlier, physical characteristics have nothing to do with whether an entity is living or not, and what we are actually doing is confusing 'intelligence' with 'life'.

If a machine can mimic exactly every function of some form of living creature, is that machine living or not? The Turing test has a very specific interpretation - it tells us when we can no longer distinguish between 'a man' and 'a machine', but nothing about 'life', and specifically not whether such a machine is alive.

We are still a very long way from developing a machine ready to face the Turing test, but there is a general impression nonetheless, that when we are, it is an indication of 'life'. The Turing test, however, was proposed purely in the context of measuring the evolutionary progress of artificial intelligence, and nothing more than that.

Nevertheless, it will always be a matter of interpretation, and even when Turing proposed his test, questions were being raised as to whether such a 'successful' machine, if or when it became reality, should be accorded 'human rights'.

The whole situation is a classic example of how easy it is to focus on 'obscure implications' of distant possibilities, rather than the practical realities of intervening events; and that reality today, is that the whole question of 'artificial life' is still in it's infancy.

The real focus of attention, however, is on artificial intelligence, and how the rapidly advancing potentials of that can be constructively applied in practical ways, without attempting to combine these with humanoid characteristics.

'Life' per se is in many respects largely irrelevant, at least for the foreseeable future, for it is the 'functions' which machines can fulfill, that matter; and if (eventually) these come to embrace most things that man can do, whether such a machine is alive or not is unimportant.

Artificial life and artificial intelligence share a great deal of common ground, but for convenience it is usual to treat them as separate disciplines, which reflect the two very different aspects of any 'functional entity' - the 'object itself', and the 'internal processes' which enable it to function ('hardware' and 'software').

Thus, artificial Life is concerned with 'complete biological entities', and studies these in terms of computer simulation and robotics[26]; while Artificial Intelligence, by contrast, is concerned with software, and the design and implementation of computer programs, which underpin 'simulation', and put 'life' into 'living'.

More recently, a further category has been added, which now differentiates three different aspects to artificial life: Hardware (living objects), Software (instructions) and Wet (Biochemistry).

The term 'Artificial life' (Alife) can also be used in the context of a discipline, to describe that 'field of study' which looks at how to explore these various features by simulation; while artificial intelligence is concerned only with 'soft' aspects of life, i.e. designing computer programs, but nevertheless, in practical terms, where most of the real progress is taking place today.

By contrast, the study of artificial life itself is still in it's early conceptual stages, with a large theoretical emphasis; also, unlike traditional modeling, which usually stresses 'major properties', work in this field concentrates on simulating the most 'simple' and 'general' features, which can then be adapted upwards later.

None of these, by any standards, are considered to be living, and aside from the definitions noted above, life can also be categorized at two different 'levels' - Strong, which has been described as trying to synthesize, rather than simulating life; and Weak, the more conventional, which still relies on simulation, rather than trying to create life outside of the test tube.

In practice, however, things are rarely that clear cut, and life can be studied by a number of different techniques, including cellular automata, particularly in the early stages; and neural networks, which indirectly can give insight into 'learning behavior', though as a technique it is more usually employed in relation to artificial intelligence.

6.8 Intelligence:

Intelligence is not the prerogative of human life. All mammals are intelligent, to a greater or lesser extent, and though it is not easy to quantify behaviour, some species (e.g. porpoises) are clearly more intelligent than others. Many animals in the category of 'family pets', can also display highly intelligent (including learned) behaviour, especially with respect to their owners.

How much, or to what extent, intelligence is possessed by lesser species, is unclear, especially in respect of whether isolated examples are necessarily typical of a species. It is difficult to devise tests or criteria, but some aspects of 'learned' behaviour (e.g. in response to natural exposure, 'learning by experience') can be interpreted as intelligent, and on that basis, certain plants have shown evidence of intelligent behaviour.

Against that background, it is not surprising that the concept of intelligence is open to interpretation, and can be defined in many different ways. Wikipedia, for example, list 10 alternatives definitions, ranging from four words, to a 9 line paragraph[5].

In human beings, intelligence is usually taken to embrace a range of abstract properties - logic, understanding, awareness - among many other, such that any complete definition would essentially amount to a list.

Nevertheless, intelligence is a faculty familiar enough to most people to obviate the need for a formal definition, and might be a suitable word to exemplify a proposal that 'meaning' and 'implication' can be conveyed clearly, and more succinctly, by 'sentiment' rather than 'literalism'. Thus, for intelligence, this might be:

> The innate ability to maximize 'common sense' at all levels of cognitive (and physiological) behavior.

That would be in the context of an 'outside', as apposed to 'personal' opinion, and the qualification in parentheses is to acknowledge the possibility of intelligent behaviour in plants[6].

Intelligence can also be thought of as a function rather than a 'faculty', and is usually manifest by a range of interrelated functions (memory, learning, understanding etc). The more extensive the range and the more effectively each is used, the higher the level of innate intelligence, and it is the collective ability with respect to all of these, which determines overall intelligence status, or quantitatively, the IQ.

So far as the evolution of intelligence is concerned, it can only be inferred indirectly, from other changes (in the fossil record) which are consequences of it.

Clearly the functions it subserved would have had to be sufficiently developed to implement the advantages which intelligence bestowed. Hence, that must have pre-dated the use of tools, for example, which required insight and implied purpose, and we know from the fossil record that stone tools were in use as early as 2-3 million years ago.

If we bring plants into the picture, these appeared some 200 million years earlier, and in some instances showed evidence of learning behaviour with respect to changing environmental conditions. Nevertheless, unlike animals, they possess neither brain or neuronal networks.

That has very significant implications, for no matter how different the manifestations, it suggests that a fundamental property of nature can seemingly evolve in two completely different ways; yet so great are the differences, physiological and anatomical, that we have to question whether this is a valid comparison at all. Especially as evolutionary sophistication of neuronal networks, is the essential physical basis for those abstract faculties of the mind, which define human intellect.

Otherwise, it would imply an intrinsic difference with respect to a common property, which would be difficult to reconcile with well established principles of evolution..

There is no such ambiguity at the forefront of evolution, however, where human intelligence is the single most important, and highly developed faculty that man possesses; it not only distinguishing human beings from most other life forms, but also provides the detached objectivity of species dominance.

Although we cannot date the onset of intelligence, which in real terms is unimportant, we can certainly identify trends, over sufficiently long intervals to display change, and if we combine these with extrapolation based on present progress, construct a simple historical projection:

Time:	Intelligence:
2-3 million years:	First evidence of intelligence (tools).
1 million years:	Useful intelligence.
500,000 years:	Advanced intelligence.
250,000 years	Human intelligence.
50 years ago	Artificial intelligence evolving.
2029 - projection:	Machines achieve precessing power of the brain.
2045 - projection:	Machines able to improve their own performance.

This is clearly a progressive pattern, and even allowing for approximation, shows an overall trend consistent with Moore's law - namely, that artificial intelligence seems destined to cover in about 90 years, much the same ground that took human intelligence over 250,000 years, or approximately 3000 times faster than organic evolution.

This tells us little we don't know already, but there is always a pertinence in visual information, which emphasizes 'patterns' more effectively than other ways, and in this case highlighting man's vulnerability - that in the foreseeable future, artificial life and intelligence will progressively take over, and eventually threaten man's position as the dominant life form; for the reality is that inorganic existence (AI), can offer a better, more secure future, than would ever be possible under the auspices of 'organic man' (arrogant, opinionated, self-destructive).

6.9 Knowledge:

This is another area of abstract thinking, which has been a source of confusion and debate among philosophers over the past two millennia, ever since Plato laid down conditions which had to be fulfilled, before any form of statement could be classed as knowledge.

Nevertheless, for present purposes, all we need is an understanding of what knowledge is and means, and we can ignore Plato's guide lines at this stage. A number of other concepts are involved, however (information, cognition, facts) each in a sense qualifying the other, and we can show how these three are related by a common definition:

> Knowledge is the possession of facts; cognitive function is how
> the mind deals with (processes) that information, and intelligence
> is how well it does it.

'Knowledge' and 'intelligence' are thus inseparable, for it is only in terms of 'knowledge', that 'Intelligence' can manifest it's abilities at all; and how this triad of functions integrates and behaves, is an important aspect of human personality.

Man's skills and technology have been advancing rapidly over the past 50 years, passing insidiously through ever more advanced staged - collecting, storage, processing - to full responsibility and eventually automation.

Progress is now bordering on exponential, with the technology of computing hardware and integrated circuits doubling every two years (Moore's law), and the ultimate objective of independent artificial intelligence probably achievable, though not in the foreseeable future.

Whatever the long term outcome, however, there will always be two major concerns:

(1) How much autonomy will these machines eventually have? In theory, as the designers, man will always retain overall control - unless or until, he decides otherwise. There will always be skeptics, however, who

envisage complacency and 'clever machines', able to by-pass safe guards and take over control (15.3).

(2) Where and How can we store so much information (increasing exponentially)? There are three essential requirements - convenience, accessibly and security - but these are not necessarily mutually compatible, and there will always be the dilemma of which one, and to what extent, compromise is allowed; in addition, from a human stand point, priorities will differ widely, depending on circumstances, and no rigid protocol will ever be practical.

At the present time, although storage technology is still theoretically adequate, the practicalities of availability are already problematic, and the reality is, that sooner or later, there will not be enough space (in whatever form) as quantity continues to increase. For the time being, Moore's law provides guidance for planning ahead, but it can only be for the interim before progress approaches exponential, and present facilities will no longer be adequate.

We are already running out of space and methods, on the physical side, while the practicalities of implementation, will soon present even greater problems, and what man needs urgently, is an alternative approach to the storage and handling knowledge,

Information Technology, however, is now slowly catching-on, and many changes are already taking place with respect to 'storage and access', though still on a relatively limited scale. The concept of a database is being revised, and superceded by knowledge base (KB), which will need data to be 'structured', with 'pointers' instead of numbers and tables. The ideal would be an 'object model' (ontology), subdivided into various different classes.

On an even wider scale, Knowledge Management envisages something much more comprehensive than handling raw data on it's own - storing and handling 'knowledge' itself. This will require entirely new method of storage and containment, and research along these lines is still in its infancy.

Purpose made tools and other equipment too, will all have to be adapted, but hopefully rather less complicated than the 'Decommissioning and Deactivating Knowledge Management Information tool' (D&D KM-IT)[7], built as part of a research project, by the D&D user community.

There is no shortage of interest in ideas such as these, but little appreciation of urgency, while the financial implications of R & D, on which ultimately the future will depend, are not even in the pipe line.

The reasons, unfortunately are not difficult to see, just difficult to do anything about. A project of this sort requires large scale international cooperation, and the one thing that puts that on semi- permanently hold, is the ties between governments and the needs of secrecy and military security. These would all have to be dismantled and moved elsewhere, before work could even begin, which might take many years to accomplish, even assuming a willingness to do it at all.

How man handles the 'knowledge crisis' which is imminent, over the next ten years, in the context of technology translating into 'practical benefits', could well determine the future of technology itself for a very long time to come.

6.10 Death:

Death in it's usual context applies only to organic life forms, and may be defined as 'the permanent cessation of life processes', after which the object concerned is defined to be 'dead'.

When this occurs as a natural event, it is preceded by the progressive reduction of those function which are considered essential to life (dying). This process in itself, is usually a matter of arbitrary definition, and can even last for years.

Outside the context of organic life, however', inanimate or non-organic objects are often referred to as 'dead', while any sort of 'planned activities' which are 'winding down', or just not going to happen (e.g. political), can be described as 'dying' or 'dead'.

As in many other circumstances, however, there are always grey area - viruses, for example, do not metabolism, do not possess cell structure (usually regarded as the basic unit of life), and therefore cannot reproduce by cell division; yet they proliferate rapidly by other means, and are one of the commonest species of organism on earth. Whether they are 'living' or 'dead' however is still a matter of debate.

The eventual prospects for artificial life is one of the great uncertainties of the future, but there can be little doubt that sooner or later we are going to have to take a very close look at what we really mean by 'life' and 'living', and the implications of that with respect to the machines which we are planning and building for the long-term future, for once artificial intelligence is able to pass the Turing Test†, and is therefore (by definition) indistinguishable form a 'living person', we have to ask ourselves one important question - "does it really matter"?

Clearly there are very significant differences between the two, which we would already be familiar with, and hence the implications behind the Turing test are not so much whether or not it confirms these differences, but whether solely from the answers given, we could distinguish between a living entity and an inanimate machine, i.e. in effect between 'living' and 'dead', while ignoring the anomaly of associating 'death' with 'functional activity'.

If this seems to present a difficulty, one possible answer is to regard death as a property of life (as a matter of definition) albeit at the extreme end of the range, but nevertheless, with the same legitimacy that entitles us to regard 'zero' (nothing) as a real number.

It would be much too simplistic, however, to extrapolate from that, to equate 'man' with 'life' (as though they were synonymous), for 'human' life is a very special sub-category of 'life', with deeply rooted abstract qualities that distinguish it from man's immediate predecessor in 'biological complexity' - the animals. This can perhaps best be exemplified as the ability to distinguish between 'right' and 'wrong' - 'consequence' from 'responsibility' - and the many ramifications of these, which constitute 'personality'- the ultimate uniqueness differentiating man from all other forms of life.

Whether an analogous distinction would ever have to be made between 'man' and 'a machine', is impossible to say at this stage. We do not even know how human beings came to acquired such abstract faculties in the first place, and the insidious continuity of evolution, from physiologically identical ancestors, offer no explanation. We can only assume, therefore, the influence of 'some other factors' which do not belong to the reality of this physical existence, but seemingly can influence events and entities which do.

It is the implications of these which make the long term future so uncertain, but eventually, with 'preservation of self-existence' a primary requirement for life, it is likely that death will eventually be abolished in some way or other (15.8). In the meantime, however, artificial intelligence will continue to evolve many times more rapidly than man, progressively

---

†   Turing Test: that it is no longer possible to distinguish between a 'man' and a 'machine', from the answers given by each to identical questions.

assuming a dominance which was never visualized at the outset, and it could well be, that long before the conquest of death, the human race itself will have been replaced by 'intelligent technology'(15.3); and if there are life forms elsewhere in the universe, comparable to our own, these too may be similarly destined, if indeed they have not already been replaced.

We look in detail at the implications of Fermi's paradox in later chapters, but it may well be, because of the evolutionary imbalance between artificial and carbon-based life, that the eventual destiny of the universe will be artificial rather than biological (15.8), with the role of the latter, in the greater scheme of things, simply to initiate the former!

With hindsight, based on the above scenario, one could postulate a universe, in which the sole purpose of organic life was to initiating artificial technology, but little else. While death existed solely to allow for the replenishment of youth, and with it the continued exponential evolution of intelligence'; whereas the alternative, a fixed population of 'eternal life', would have embraced declining intellect, dementia, stagnation and regression.

Reference (Chapter 6):

1. Wikipedia article (14 September 2013), Life.
2. Wikipedia article (3 December 2013), Phylogenetic Tree.
3. Wikipedia article (5 May 2014), Evolution.
4. Wikipedia article (10 March 13), Human Evolution.
5. Wikipedia article (15 September 2014), Intelligence.
6. A. Trewavas, Green plants as intelligent organisms (Trends in Plant Science 10 (9):2005).
   (Click on this reference in Wikipedia to access original).
7. Wikipedia article (22 April 2013), D&D KM-IT.

Chapter 7

REQUIREMENTS FOR LIFE

There are a large number of pre-conditions which have to be fulfilled for life to be possible at all. In many respects these might be regarded primarily as 'environmental', but individually they can apply to a variety of different, but related features, with respect to the universe as a whole, a planetary system, and the bio-physical properties of any potentially suitable planet, such as earth.

7.1 The Universe:

The age of the universe depends on it's geometry(2.3), and this was predetermined at the time of the Big Bang singularity (3.1). Theoretically, a universe can have any age, but for life to be possible, it has to be at least old enough to allow sufficient time for the necessary chemical elements to build up and become available.

Very few of these are formed at the time of creation. Primordial nucleosynthesis (4.3), for example, lasts only a few minutes, and produces (for all practical purposes) only pure hydrogen and helium, out of which star themselves are formed later (4.8).

Stars first appeared between 400 - 600 million years after creation, and the heavier elements (all elements other then H and He) were then synthesized in the cores of first generation stars (secondary nucleosynthesis). Only when these eventually died, and disseminated their contents into space, were the elements then freely available, and life could begin to form.

If the universe had been only one tenth of it's present age, none of this could have happened, and even the solar system itself would not yet have formed.

Had the universe been ten times older, however, most stars would have moved off the main sequence, to become white dwarfs, while planetary systems would long since have come and gone.

Hence, biological factors required a 'window' in the lifetime of any universe - a "golden age", not too young and not too old, for the necessary chemicals to build up before life itself could come into being.

7.2. Geometry:

We saw in chapter 2 that theoretically there are only three possible geometries which a Friedman universe can have - hyperbolic, spherical and flat (Figure 2.2).

These represent solutions to Einstein's equations of general relativity, but in practice any real universe must either be 'open' or 'closed'. 'Flat' is a purely theoretical demarcation and in practice, such a universe could never exist. It would be far too unstable, and very soon revert to one or other of these options.

The final geometry would normally be decided at the outset, when the curvature constant is fixed at the time of creation, Thereafter, it's numerical value can change, but not the sign.

It is one of the mysteries of cosmology that this did not happen, and the present universe has existed 'on a knife edge', balanced precariously between 'open' and 'closed' for over 13 billion years.

How long it can remain stable we have no way of knowing, but after such an interval and against such odd, it could very well be permanent.

Had the geometry been different from the outset, the history of the universe would have been very different too, for either way it would have run it's course and ceased to exist long ago, and life, which took 9 billion years to appeared, might never have done so, or at the very least would not have had time to evolve.

7.3 Fine Tuning:

There are approximately 25 dimensionless physical constants; this number is not fixed, and others could still be added. They are critically important to the functioning of the universe, cosmic homeostasis, and the biochemistry of life.

Their cardinal characteristic is fixed values throughout the observable universe, which do not change over time†, and (in most cases) a very narrow range of values with respect to the function they fulfill.

It is this latter property which interests cosmologists, for these critical values appear to be pre-requisites for the existence of life; had they been different, by even the smallest amount, the physical properties of the universe would have been different, and organic life would not have been able to form.

The fine tuned universe is an accepted concept in modern cosmology: the facts themselves are not in disputed, but there is wide disagreement regarding their interpretation.

Stephen Hawking puts it quite succinctly:

> The laws of science, as we know them at present, contain many fundamental numbers, like the size of the electric charge of the electron and the ratio of the masses of the proton and the electron. .... The remarkable fact is that the values of these numbers seem to have been very finely adjusted to make possible the development of life.

For the present, how we approach the universe really depends on which we regard as most likely - coincidence, some other logical interpretation based on known science, or as Hawking would seem to

---

† The temporal stability of these 'constants' is a long standing topic of debate, and some recent work in Australia (as yet unconfirmed) suggests the fine-structure constant may have alter by a tiny amount since the universe was formed 13 billion years ago.

imply, a relationship more suggestive of 'intention' - i.e. 'pre-ordained' (chosen) and set in advance precisely so that life could arise. That would certainly be an extreme and very much minority view, though essentially little different from the strong Anthropic Principle (10.2), which embraces both 'purpose' and 'fulfilment'.

There are many 'lesser' interpretations, however, while others dispute the whole concept of 'fine tuning' altogether. Their arguments range from 'alternative biochemistry'; that other forms of non-carbon based life may be possible (14.5), especially when the whole universe is taken into account; that sufficient allowance is not made for the potential adaptations of evolution, or that 'fine tuning' is simply anthropomorphism applied to physical constants.

The most logical argument, however, disregards the physical variables altogether, and simply proposes that life will arises where the environment is most suited for it to do so. Implicit in this, however, is that there must be alternative options to choose from - i.e. other universe, such as a multiplex (3.6), and indeed the possible existence of these is now something which is beginning to be taken seriously (e.g. Brane cosmology, 1.8).

The other major alternative to all of these implies 'choice', and that the universe was 'pre-ordained' so that life could arise, and eventually progress to sentience and intelligence. Inevitably, however, any suggestion of 'higher intelligence' (3.8) soon gets bogged down with the difficulties of explaining 'where' or 'what' such an entity might be, or blurs into 'religion', while most scientists have an instinctive aversion to invoking God, in whatever form, into their professional lives.

Nevertheless, this is not the first time such a suggestion has been seriously proposed. The original Copenhagen Interpretation of quantum mechanics in the mid 1920s, required an 'outside observer' to make it work (1.3); as a theory it was soon revised and replaced, but surprisingly little objection to it at the time, based on these grounds, as opposed to others.

7.4 Parent Star:

The Physical characteristics of a parent star are important to planet habitability, though it's consequences for a given planet will depend on proximity, and therefore the nature of that planet's orbit. Low eccentricity is essential for supporting life, and the earth's orbit is very close to circular.

Spectral class determines surface temperature and mass, and our sun (G2, Main Sequence) has a number of important characteristics in relation to habitability and life:

- A significant life span (several billions years), which allows time for life to evolve.
- An ideal UV spectrum of high frequency radiation.
- Allows for the presence of liquid water on planets orbiting at optimal distances for supporting life.

Unfortunately such stars are not common - only between 5 and 10% in the local galaxy, while over 70% of stars in the Milky Way are red dwarfs. These are usually considered too small and cool to support habitable planets, though they would have two big advantages otherwise - extremely common and very long lived.

However, a recent study has suggested that cooler stars may actually host warmer planets, and whether red dwarfs might be suitable after all is now one of the most important questions in the whole field of planetary habitability.

Suitable habitable zones and stable luminosity are also important, as is metallicity, because a high proportion of heavier elements in the protoplanetary disc makes planet formation much more likely.

7.5 Circumsolar Habitable Zone[1](CHZ):

This was discussed earlier in chapter 5.10. Water is essential to life, and the CHZ defines that region of space around the sun, in which a planets can possess a suitable atmosphere to support liquid surface water.

Zones can be estimated in a number of ways, and for the solar system, a conservative CHZ lies between about 0.725 and 1.4 AU out from the sun. Although zones are usually slightly asymmetrical, the earth's orbit lies close to the center of this one.

Other factors to be taken into account include gravity, and the 'greenhouse' gas CO_2, but if the earth were either 'closer' or 'further' from the sun, by as little as 5% (4 million miles), the necessary coexistence of water in it's three different 'states' might no longer be possible.

Not all planets are necessarily suited for life either, for there may well be other factors mitigating against water. For the solar system, for example, an 'extended' CHZ, outwards for up to 3.0

AU, would embrace not only Venus (just), and the Moon, but now also Mars, and the dwarf planet Ceres - none of which are known to have surface water, for a variety of different reasons†.

Further out still, and in 2011 the Cassini probe found evidence of water on the surface of two of Saturn's moons, Titan and Enceladus (which may lie within the CHZ of Saturn itself), and later led NASA scientists to claim that "Enceladus is emerging as the most habitable spot beyond Earth in the Solar System for life as we know it"[2].

Such findings lend support to Panspermia (Chapter 9).

7.6 Geology††(origin of earth's water):

A material universe requires a stable physical environment, compatible with the 'permanence of existence', and the physical and chemical interactions which that entails.

The Hadean eon is the earliest defined period of the earth's geological past, when it was forming out of the protoplanetary accretion disc, 4,600 Ma.

The early earth initially would have been a hot mass of tectonic activity, and continuous bombardment with meteorites and larger body impacts. It is thought that collision with one of these, some 100 million years later, scattered debris around the earth which later coalesced to form the moon. Once this newly formed earth- moon system settled, lunar gravitation gradually stabilized the earth's axis of rotation, and this helped to optimize an environment in which life could later develop.

Oceans are thought to have formed early in this scenario, with water at one point, covering most of the earth's surface. Over time, however, meteorite activity, and surface bombardment, gradually vaporized these, forming thick cloud layers surrounding the whole of the earth, and which would eventually form the pre-biotic atmosphere in which life would later develop.

The continental shelf probably formed later, but by 4,000 Ma, when the Hadean period ended, oceans and continents were already demarcating.

Clearly, it was very necessary that initial geological conditions should be 'right', or at least conducive for life to form, but these conditions seem

---

† Water vapor coming from the surface of Ceres (presumably evaporation from surface ice) was observed by the Hubble space telescope in January 2014.

†† Units: Ga - giga years (1000 million years ago); Ma -1 million years ago.

positively hostile, and perhaps the question we really want to answer is whether life on earth (or in the universe) came to exist because of these conditions, or in spite of, them?

Life after all begins by 'self-assembly' of individual elements - so that if the former, it would have been purely random as to where the right circumstances existed or not; while if the latter, how were these elements able to start to join together, if the conditions were not suitable in the first place?

Whichever one we choose, it is difficult to avoid the conclusion that things would have been a great deal simpler with some form of 'Outside Help'(3.8).

7.7 Elements:

The 'building blocks' of all matter. There are 98 naturally occurring element - listed sequentially by atomic number (number of protons in the nucleus) as the Periodic table - and 19 others, which claim to have been synthesized, though several remain unconfirmed.

General Properties:

- Elements are pure substances, composed of single type atoms.
- 'Normal matter' (which includes all naturally occurring elements)
- only accounts for 4% of the total matter content of the universe.
- Only one third of all elements occur individually, usually as mixtures; the remaining two thirds as compounds.
- H and He are the commonest elements in the universe; O2 the commonest in the earth's crust; and Fe the most abundant element (by mass) making up the earth.

Carbon is the basis of all known forms of life, and it's particular ability to bond with oxygen, hydrogen and nitrogen (the chemicals of life), and to form self-assembling molecules, is the basis of organic life (14.4).

It is abundant on earth (the 15th commonest element) and the 4th commonest element in the universe after H, He, and O, but nevertheless, owes it's existence to such a rare property of the carbon-12 nucleus, that the likelihood that life could have arisen by pure chance alone, is statistically close to impossible. This unexpected property is the possession of a resonance level (vibration frequency) exactly matching that of the beryllium-8 nucleus, which is normally unstable, and splits apart almost immediately after being formed.

Because of this match, however, if beryllium absorbs a third á-particle (4He) just as it is being formed (triple-alpha process), instead of disintegrating, it becomes a stable carbon-12 nucleus (3á particles) and so ensures the continuity of carbon production.

Hence, we have the following relationships:

Normally:

$$^4He + {}^4He \longrightarrow {}^8Be \text{ [disintegrates immediately}$$
$$\longrightarrow 2 \text{ á particles]}$$

With the triple á process:

$$^4He + {}^4He \longrightarrow {}^8Be \text{ [unstable]}$$
$$^8Be + {}^4He \longrightarrow {}^4He + {}^4He + {}^4He$$
$$= {}^{12}C \text{ [stable]}$$

Thereafter, all the remaining elements are synthesized by fusion within the cores of stars.

7.8 Biochemistry:

A specialized discipline which deals with the chemistry of life, and the complex processes required by the metabolism of living matter.

These processes have developed progressively, to match the requirements for describing evolving life forms at any given stage, and ultimately those of human beings - the most advanced known form of life, and pinnacle of terrestrial evolution - but elsewhere in the universe, or what the future holds, only time will tell.

Just six major elements - hydrogen, carbon, nitrogen, oxygen, calcium, and phosphorus make up 99% of human life, while the more complex bio-molecules of organic matter are usually classified into 4 main groups, largely determined by structure and function†:

---

† NOMENCLATURE::

--Monomer - a molecule that can combine chemically with other molecules (polymerization) to form a chain of molecules, known as a polymer.

-- Polymer - a macro-molecule composed of a chain of monomers.

-- Polymerization - the process of combining monomers to form polymers.

1. Carbohydrates: Composed of monomers, including glucose and sugars. They are an important source of energy, and also provide structure, e.g. cellulose, and the genetic molecule RNA, among other functions.

2. Lipids: Naturally occurring molecules, usually comprising glycerol and fatty acids (triglycerides). Their main functions are storing energy and providing structure, e.g. cell membranes.

3. Proteins: Large macro-molecules, consisting of chains of amino acids, the sequence of which (determined by genes) distinguishes one protein from another. Functions include acting as catalyst, transporting molecules, replicating DNA and responding to stimuli.

4. Nucleic Acids: large (macro-molecule) polymers, made from chains of monomers known as nucleotides. They are essential to all known forms of life, and include DNA and RNA, responsible for the processing of genetic information.

7.9 Atmosphere:

There are a number of theories regarding the composition of the earth's early atmosphere. Alexander Oparin, in 1924, was the first a propose a reducing atmosphere (8.3) - hydrogen, methane and ammonia. He argued that little free oxygen was present when the earth was formed, and that it could have prevented the synthesis of some of the necessary building blocks for the later evolution of life[4].

This idea was first taken up by Stanley Miller and colleagues in 1953, who showed in a number of experiments (8.4), that amino acids (the building blocks of protein) could be created by passing an electric discharge through a mixtures of hydrogen, methane and ammonia gases, together with water vapor. These experiments were later repeated for different compositions of gases, and with different forms of energy, including lightening, ionizing radiations (radioactive potassium), UV radiation and electron beams. Organic compounds were formed with all of these, but none if oxygen was present[5]. There was also geological support for these early conditions, when large amounts of ferrous (reduced) iron were found embedded in some of the earliest pre-Cambrian rock strata[5].

Many studies along these lines have been carried out since, but none have progressed beyond amino acids and polypeptides, and opinions today are moving away from an early reducing atmosphere composed of hydrides. It is unlikely, for example, that the earth's gravity would have been sufficient to hold on to it's molecular hydrogen, and much of this may have seeped back into space, probably quite early on.

We also now have a better understanding of the earth's early volcanic activity, influencing many variables (temperature, gases, water vapor, carbon dioxide, etc), and the extent to which it's surface would have been exposed to almost constant meteorite bombardment[6]. This would have drastically altered the environment, vaporizing oceans (which probably formed early) and rocks, to form high altitude clouds completely surrounding the planet. Eventually these would condense, and heavy rain falling over thousands of years would gradual restore the surface water, to form seas and oceans.

Under these conditions, it is thought that the early atmosphere was probably very different from the above - only about 60% hydrogen, with now up to 20% oxygen (mostly as water vapor), 10% carbon dioxide, and smaller amounts of nitrogen, free hydrogen, methane and inert gases.

Oxygen was produced by photosynthetic cyanobacteria, from about 2.4 Ga (from today) onwards, and initially it was removed by organic matter; when that became saturated, however, after a few hundred million years, free oxygen built up in the atmosphere and eventually wiped out most of the earth's population of anaerobic organisms - the Great Oxygen Event, and one of the most significant mass extinctions of all time (6.5). Oxidation of methane also triggered the Huronian glaciation, lasting 300 million years - almost half as long as life itself - and one of the most severe ice ages in history.

We look at the relationship between these early atmospheres, and the origin and evolution of organic life, in chapter 8.

7.10 Energy:

This was an immensely active period in earth's formation, with ongoing geological and atmospheric upheaval, together with continuous surface activity, from constant meteorite bombardment to asteroid-sized body impacts, involving energy in many different forms:

1. Heat in some form or other was common to all, and initially would have supplemented that residual from the protoplanetary disc. A less common source, no longer significant today, could have been

'gravitational sorting', from convection within molten material (still present between the earth's core and the mantel).

2. Solar Energy. This is by far the dominant source. Each year 260,000 cals of radiant energy fall on every square cm of the earth's atmosphere - enough to boil away a layer of water 12 feet deep, covering the whole of the earth's surface[7].

Light energy, however, can only bring about chemical reaction if it gets absorbed by at least one of the components of the reacting mixture, and for solar radiation, most energy lies in the visible region which is transmitted by atmospheric gases.

Only the UV region, therefore (where energy falls off rapidly) would have been available for the synthesis of organic compounds, so we have no way of knowing what that precise amount of energy might have been. Nevertheless, had a chemical such as formaldehyde been present in the atmosphere (even though we have no reason to suppose that it was) much larger amounts of UV radiation would have been absorbed.

This possibility was investigated by a team of Indian microbiologists, primarily concerned with aspects of abiogeneses (8.2) and microscopic particles known as Jeewanu (Hindu for 'particles of life'[8]), but findings were inconclusive.

3. Lightning and Electrical Storms take place close to the surface, where solar energy would largely have been filtered out by overlying cloud layers. Some authorities have drawn a parallel between frequent electric storms observed in the atmosphere of Jupiter, and the reducing atmosphere on earth. Such activity, being more confined, could have been influential in the formation of organic molecules, or in providing the focal energy, perhaps necessary to progress monomers to polymers, though this has not yet been established in the laboratory.

4. Volcanic Activity. There was no shortage of this in the early Hadean period of the earth's formation, though much of the energy generated would have been heat, and therefore dissipated underground. One useful thing about volcanoes is that they are still present today, where almost identical environmental conditions can usually be studied or simulated.

Sidney Fox[9], for example, collected volcanic material from a cinder cone in Hawaii. He found that temperatures over 100 °C were present within 4 inches of the surface, and postulated that this might be the sort of environment in which early life could have formed, and later washed out, together with volcanic ash, into the sea.. Worldwide, many different forms of volcanic activities can be studied, with respect to ejecta, lava flows, escaping radiation etc, and in particular the wider environmental consequences, which would have been at least as important then as they are today.

5. Hot Springs. These would also have been widespread in the turbulent volcanic activity of the early earth, and an additional source of local heat and energy. Many pre-biotic chemists, however, see their contribution as insignificant, by companion with volcanic activity.

6. Shock Waves. Most of these were probably the consequences of large meteorites traversing the atmosphere. It has been calculated that an object 5 miles in diameter, traveling at 5 mps, could have generated as much as 1012 tons of organic matter in it's wake - a very substantial contribution, even for comparatively rare events[7]. This would then have sedimented on to the surface of the pre-biotic environment, and probably had an adverse effect, by stifling activity or reducing the incidence of radiant energy.

There would always be some degree of interaction between the differing roles of 'environment' and 'energy sources', but the overall balance would always be positive, with respect to the needs of evolving life at any given time.

7. Radioactivity. This would be widespread in the earth's crust, and an important potential source of energy[7]. A principle source today (among others) is potassium 40. From it's known half-life, and the amounts present today, we can extrapolate backwards to calculate the energy available from this source in the pre-biotic era, and that turns out to be of the order of 12 x 1019 calories.

This is only about one thirtieth of the energy we get from electromagnetic radiation, but it's significance lay as much in 'availability' (to penetrate) as in 'quantity'. Others, however, argue that radiation of any sort would more likely be harmful than beneficial to incipient life, and we will probably never know what the relative importance of the different energy sources was.

These are more than enough sources of energy, but they would not all have been used concurrently; more likely they would have been accessed opportunistically, i.e. each in proportion to the extent to which it was the most available or convenient form. Variety was essential, however, given the variations in energy availability, to ensure the continuity of any evolving biosphere.

7.11 Security and Risks - how safe is life?

This will always be a balance between many different facts and circumstances,

Life evolves so slowly, over such long periods of time, that evolution is imperceptible; something we instinctively take for granted but are not consciously aware of, as opposed to the progress and change taking place in every day life, which embrace practical realites - the safety and security of existence, the permanence of continuity, and more often than not, the fallacy of correlating 'future' with 'past'.

As a general rule, past events have little predictive value, and offers no guarantee of future circumstances - either with respect to safety and security, risks and hazzards, or the stability of evolution itself.

Both the universe and life stem from backgrounds of extreme improbability; by any standards we should not be here at all, and there are aspects of these which are still potentially relevant today - unstable geometry (2.3, 7.2), and certain aspects of 'fine tuning' (7.3) - but impossible to assess in terms of 'future risk', when they already conflict with logic and laws of probability..

We might add Carbon Resonance † (7.6) to that list, as something which could terminate organic life entirely, should it ever change in the future (however unlikely); and although small amounts of carbon are produced elsewhere, they could never be sufficient to meet the needs of organic life, as it now exists.

Cosmic Rays in interstellar space can reach energies as high as 3×10^{20} eV (equivalent to the energy of the universe within a millisecond of the Big Bang). The earth, however, is protected from radiation of this sort in a number of ways: mainly by it's atmosphere, which is opaque to primary radiation, while secondary cosmic rays, which can still reach the

---

† A specific resonance level (vibration frequency) of an excited carbon-12 nucleus, which makes possible the triple alpha process, and the formation of carbon out of helium.

surface, are attenuated by atmospheric absorption. In addition, both the earth's magnetic field, and the interplanetary magnetic fields, embedded within the solar wind, deflect cosmic rays.

Butterfly effects, by contrast, are very real, but cannot be detected in advance. Tiny and indiscernible initially, and usually unsuspected, they grow and enlarge insidious, until eventually producing an effects, which by that time might be difficult to counter.

Asteroid impact is not only the most likely threat, certainly in the short term, but also the only 100% predictable risks, as well as eventually preventable - once we have the necessary technology.

An object between 2-3 miles in diameter would release an amount of energy equivalent to simultaneous detonation of several million hydrogen bombs. By any standards, extremely uncommon, yet the Tunguska meteorite in 1908, was almost twice that size, and only the fact that it disintegrated in the lower atmosphere, rather than on impact, prevented a catastrophe. As it was, over 80 million trees are estimated to have been destroyed.

Nevertheless, over geological periods of time, impact events have played a significant role in the formation of the solar ssystem, but had any of these occurred in modern times, we would certainly not be here today.

Every planet in the solar system has been bombarded over the ages, and the surfaces of those which lacked the protection of an atmospheric are peppered with impact crater.

The geology of the moon's surface has provided the best source of information, from which we can extrapolate, and it is surprising how much can be deduced indirectly from these, with respect to craters on other planets.

The size of an impact craters correlates with the size of the asteroid causing it, and from that we have been able to estimate 'frequency-size ratios' for asteroid activity in the geological past.

An asteroid about half a mile in diameter will impact the earth roughly once in every 500,000 years; 3 miles across, once in every 20 million years; and the last known impact for an object of about 6 miles in diameter, was 66 million years ago (Gulf of Mexico).

Nevertheless, the consequences of even a small asteroid, less than a mile across, impacting in a populated area, would be devastating, and with secondary consequence (homelessness, disease, crime, socio-economic, environmental, etc) possible as damaging in the long run as the event itself.

Impact by any object much over 6 miles in size, however, would irrevocably damage the biosphere, and certainly be classed as an 'extinction event'. Extremely rare though that might be, it is none the less a virtual certainty at some point in the future, and since it is now 60 million years from the last comparable event, that could be 'sooner' rather than 'later'; while some statisticians have even claimed that the such an event is now overdue.

Within the past couple of decades, technology has now reached a stage where it may well be possible to avoid or prevent at least some of these impact disasters. In 1992, NASA instigated the first

'Asteroid Watch' programme - initially with the aim of detect 90% of near-earth objects, larger than one kilometre, 'within 25 years of impact', and over the next few years, this 'target' was reduced to 'within 5 years of impact'.

These are huge 'widows' of advance warning, but there are many risks which can arise much more quickly - gravitational influences, for example, which could quite unexpectedly deflect an object into an 'impact' trajectory.

Today, there are numerous 'Spacewatch' projects in operation, and others nearing completion, such as ATLAS (Asteroid Terrestrial-impact Last Alert System) in Hawaii, due to become operational in 2015, and able to give a one week advanced warning for near- earth asteroids down to 150 ft in diameter.

All of these 'observing stations' rely on radar, high quality optics, and computerized 'Impact Probability' calculations, together with an increasing number of 'Space based' detectors.

The real problem, however - what to do about something when it is detected? - remains on the drawing board. There is no shortage of ideas, though where or how any of the these could realistically be evaluated, is yet another difficulty. The following have all been considered, though even this list is not exhaustive:

Nuclear explosions.
Kinetic impactors.
Gravity traction.
Rockets.

Laser ablation.

Ion beams.

Focussed solar energy.

Self Destruction through nuclear warfare is now much less of a reality than in the past, though neither 'politics' or 'leaders' can ever be predicted, any more than another Hitler can be ruled out.

However, this category of risk would now more likely be due to genuine 'mistakes', through error or ignorance, and it has been suggested, for example, that if the large Hadron Collider could ever achieve the energies of the Big Bang, this might trigger an unconstrained chain reactions, which could threaten the integrity of the entire planet.

This has always been seen as a theoretical risk, since machines of this sort were introduced, and there can be little doubt that eventually, they will be able to achieve very nearly energies of that level. Nevertheless, the physicist involved and responsible for such work, are hardly the best people to 'judge and decide', and it is not surprising that none of these, when asked, will ever admit to seeing their own work as a possible threat.

Climate change is likely to be a contentions issue for some time, not least because scientific objectivity will always conflict with commercial interests, and there can be few other areas where the eventual implications could be almost as drastic as the events themselves.

In the long term, however, past geological eras in the earth's history may eventually recycle, perhaps over equally long periods of time, and man will have to evolve and adapt, in ways which would be unrecognizable today.

Whether any of these would amount to a 'threat', however, in the present context is debatable, but given man's past and present abilities to adapt, possibly not.

Associated consequences of climate change, however, could be almost as hard to deal with as the event itself, and in particular, factors such as radiation, in relation to atmospheric changes.

Disease may be something man feels he has 'under control', but the 'flu epidemic of 1918 killed between 50 and 100 million people world wide, and lesser pandemics have occurred regularly throughout history.

Pandemics are conventionally attributed to the spread of some new strain of virus from animals to man, but we can never rule out panspermia either (9.1), and the possibility of alien micro-organisms, resistant to all known treatments, which could wipe out a substantial proportion of the world's population. Unlikely though that may seem, with the escalating resistance to antibiotics (a battle which we are not yet winning today) large-scale medical disasters will always remain one of the possible ways in which life could become extinct.

Should that eventuality ever come about, however, the next relevant issue is then 'how long before it will all start over again? - either by a suitably adapted version of abiogenesis, or perhaps a different form of organism, imported via panspermia.

In the very distant future, about 4,000 million years from now, the sun will expand to become a red giant. Probably by then man will long since have colonized other planet, and the earth, having served it's purpose as incubator, will duly be cremated.

References (Chapter 7):

1. Google, The Habitable zone.
2. R. A. Lovett, Enceladus named sweetest spot for alien life (Nature, 31 May, 2011).
3. Wikipedia article ((28 November 13), Origin of Water on Earth.
4. A. I. Oparin, The Origin of Life (New York: Dover, 1952).
5. C. Ponnamperuma, The Origin of Life (Thames & Hudson Ltd, London 1952).
6. N. H. Sleep, et al, Annihilation of ecosystems by large. asteroid impacts on early Earth (Nature 1989, 342, 139–42)
7. L. E. Orgel, The Origin of Life (John Wiley & Sons, New York, 1973).
8. M. Grote, Jeewanu, or the 'particles of life'. (J. Biosci, 2011, 36(4): 563-579) [Direct access to this reference on Wikipedia].
9. C. Ponnamperuma, The Origin of Life (Thames & Hudson Ltd, London 1952).

Chapter 8

THE ORIGIN OF LIFE[4][6][7]

(A biogenesis)

So far as we know at the present time, earth is the only planet on which life exists. We know a great deal about that, however, and by extrapolating backward, have been able to trace an orderly sequence of evolution (6.4) to the earliest evidence of fossil microbes, 3,700 million years ago, i.e. to within 1000 million years of the earth itself forming, out of the prot-planetary disc (6.4).

Whether life originated in the pre-biotic conditions prevailing earlier than that (8.3), or was 'seeded' from elsewhere within the universe, we do not know; but either way, given the enormous size and complexity of the universe, it would be surprising if earth was the only inhabited planet.

Nevertheless, we have been searching (unsuccessfully) for evidence of life elsewhere, for almost 50 years, and while there are many reasons which might account for that (Chapter 15), instinct is a compelling incentive, and that leads us to look seriously at the potentials implicit in panspermia (chapter 9).

8.1 Life in the Universe:

Within the past 4 years, the Kepler space observatory(12.9) has revolutionized our understanding of stellar and planetary characteristics, and entirely changed our thinking about the feasibility of life elsewhere. Red dwarfs are the commonest star type in this galaxy, but planetary systems are extremely common, many with earth-type planets, and many of these within the habitable zones of their parent star.

The latest estimates place the number of exoplanets at between 100 and 400 billion, in our galaxy alone, and so far as more detailed findings are concerned, as of February 2014, Kepler observations had confirmed over 900 exoplanets, associated with more that 76 stellar systems, and a further 2,903 unconfirmed planet candidates.

These are impressive figures, and they certainly heighten the anomaly which the Fermi paradox seems to presents (chapter 11); but they are non-specific, nevertheless, and there are many other factors which might influence the presence of life.

As things stand, however, they are very much in keeping with panspermia, and if we make a closer analysis, the case for that becomes overwhelming: present evidence suggests that life on earth began about 3.7 billion years ago, with the oldest fossil record dating from 3.4 billion years. Hence, if we assume that biological complexity increased exponentially throughout evolution (effectively a biological application of Moore's law), then life in the universe could have begun almost 10 billion years ago - long before the earth even existed, suggesting that life must have been 'seeded' to earth billions of years after it first appeared elsewhere in the universe.

The difficulty with these figures, however, as they stand, is that they go back almost to the Big Bang, and certainly to a universe in which formed structure would not yet have appeared. In general, they are more indicative of an alternative cosmology altogether, without a Big Bang; or even pre-dating the Big Bang.

Assuming these difficulties could be avoided, however, Panspermia (in any form) has a number of corollaries:

- Origin still needs to be explained, irrespective of where or when it took place
- Young 'seeds', hibernating over very long travel times, could be (genetically) damaged, or even 'out of date' on arrival.
- Life would still have to be carbon based, irrespective of origins,

- Conditions would have had to be analogous to 'prebiotic' earth.
- Panspermia may be suggestive, but it is not synonymous with extraterrestrial life, which requires 'identification' first.
- Panspermia has no correlation with the nature or consequences of evolution; these can only be judged by the outcome.
- We can draw no conclusions regarding alternative life forms, including artificial life, other than by direct contact.

8.2 Abiogenesis:

There can be no doubt that life must have arisen in the first instance from non-living matter[1], but how that came about is not known, and initially, spontaneous generation seemed the only way to explain the rapid growth and proliferation which microorganisms display.

Nevertheless, that was not really an 'explanation' in any true sense, without some from of 'mechanism' to describe how it took place.

In the meantime, the only alternative was that life must have arisen from pre-existing life, and this became known as biogenic, and the actual process by which it took place, as reproduction - now regarded as essential, and retained as one of the primary criteria for any definition of life, ever since.

As an explanation, however, 'biogenic' was no better than it's predecessor, and though it removed reference to 'inorganic matter', it was primarily a theory of 'perpetuation', said nothing about 'origin' itself, and left the question of 'first life' (together with that of 'first cause' - how the universe itself began) as a purely philosophical issue. Both have been debated as such by scholars and theologians, since the time of Aristotle, and his concept of 'infinite causal regression'.

As a theory, Spontaneous generation lacked supporting evidence, was only introduced because there was no better alternative, and it was strongly opposed by many eminent figures, including Darwin and Pasteur. Indeed Pasteur described spontaneous creation as a "dream ", and subsequently with the qualification that "life only arises forms other life forms".

It was Darwin, however, in 1871, following his voyages in the Beagle, during which he encountered a great variety of environmental condition, who suggested in a letter to a friend, that life may have arisen in some:

"warm little pond, with all sorts of ammonia and phosphoric salts, lights, heat, electricity, etc. present, so that a protein compound was chemically formed ready to undergo still more complex changes"[2].

Surprisingly, this aroused little interest at the time, and it was almost 50 years before the question of how life began, became topical again. This was in 1922, when the Russian biochemist, Alexander Oparin, became interested in the earth's early atmosphere, and proposed an alternative theory of spontaneous generation, to explain how life might have originated.

8.3 Oparin's Theory of Life[3]:

Spontaneous generation, in it's original form had been opposed by scientists such as Pasteur, on the grounds that in an oxygen rich atmosphere, such as the earth has today, preexisting organisms would immediately consume any spontaneously generated organism, and therefore no 'chain' of propagation would be possible.

Oparin, however, argued that present conditions were very different from those in the past, where the initial atmosphere of the primordial earth, would have been strongly reducing(7.9), and that under these conditions spontaneous generation could have occurred. He pointed out that the earth's early atmosphere would have consisted largely of hydrogen, from the proto-planetary disc out of which the earth was forming, together with hydrides - methane, ammonia and water vapour - such as those found in the atmospheres of the gas giants. He saw these as the raw material of life, and argued that because there was essentially no difference in composition between living organisms and lifeless matter, the characteristics of life must have been a consequence of the way in which matter itself evolved.

The surface of the early earth initially would have been largely a solution of inorganic matter, but under the influence of sunlight, able to filter through the reducing atmosphere, organic molecules would form and gradually build up into a solution of proto-organic, organic and inorganic molecules (pre-biotic soup). A similar suggestion was also made around the same time, by the geneticist and evolutionary biologist J. B. S. Haldane, and generally shares credit with Oparin, for these ideas.

These structures grew and combined into ever more complex forms, together with new properties, through interactions between molecules, to eventually form localized macroscopic systems, or 'droplets' (coacervates). These had remarkable potentials compared with other molecular combinations, and could either 'grow', by fusing together with other droplets, or 'reproduce' by dividing into 'daughter' droplets; they also possessed primitive metabolism, in which simple 'selection' could take place, with 'favorable' factors preserved, and those less favorable, discarded.

There is no doubt that Oparin was ahead of his time with many of these ideas, but unfortunately he was not able to follow them up himself, with laboratory experiments, and as with Darwin's early suggestions, they too fell into abeyance for the time being.

In 1952, however, Stanley Miller, at the university of Chicago, became interested in Oparin's proposals, and carried out a series of experiments to test out abiogenesis, under simulated laboratory conditions.

8.4 The Miller-Urey experiment:

This was an experiment designed to test out Oparin's hypothesis that under the 'reducing atmosphere' conditions of the early earth, spontaneous chemical reactions would synthesize complex organic compounds from simpler precursors, as the preliminary stages in the evolution of life.

The apparatus (Figure 8.1) consisted of two flasks connected into a closed loop, and the whole apparatus filled with an 'artificial atmosphere' of methane, ammonia and hydrogen, together with liquid water in one of the flasks, where it could be heated to induce evaporation. The water vapour then circulated, mixed in with the 'atmospheric' gases, and passed through the second flask, where

Figure 8.1

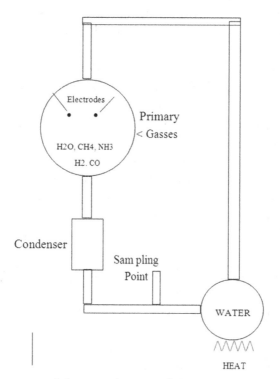

Schematic diagram of apparatus
for the Miller-Urey experiment

Steam circulates from the boiling water in the lower right flask, through the atmospheric gases (ammonia, methane, hydrogen) in the upper left flask, where the mixture is exposed to regular sparking (simulating lightning), before passing down into the condenser, where the water condenses out and continues to circulate[4].

it was exposed to sparks between two electrodes, to simulate lightening

On leaving that flask, the gases were then cooled to condense the water, which trickled back into the first flask and continued to circulate.

Within a day, the mixture had turned pink; after 2 weeks, up to 15% carbon had formed into organic compounds, 2% of which were amino acids (out of which proteins are formed), 18% of methane molecules were now bio-molecules, and the rest had become hydrocarbons; sugars were also formed.

In subsequent experiments, though most mixtures were non- racemic (14.1), in those which were, both L and R-handed optical isomers were created in equal amounts.

In 2007, vials which had been sealed and preserved from the original experiments were examined using better and more advanced analytical equipment, and now found to contain over 20 amino acids (compared with only 5 that Miller was able to identified) and indeed more than occur naturally in living tissues[5].

In addition, major volcanic activity in the early earth, which Miller had not been aware of, would have altered the composition of the early atmosphere, and when this was taken into account, and the experiment repeated with the addition of CO2, Nitrogen, hydrogen sulfide and sulfur dioxide to Miller's original mixture, many more diverse molecules were produced.

Overall, the experiments showed that with the input of energy to a suitable mixture of gases, simple organic compounds - the building blocks of proteins - and other macromolecules could be formed.

8.5 Current Theories of the Origin of Life[1][4][6] :

There is no single accepted theory for the origin of life. The consensus opinion still rests with the Oparin-Haldane hypothesis as a 'working framework', together with a large number of miscellaneous discoveries and conjectures. These usually focused on one or other of the possible mechanisms involved, and in that respect were plausible, so far as they went. However, it would be difficult to envisage a role for all of them (together), and almost impossible to specify accurately what that might be.

We can divide the Current Theory of Abiogenesis into three parts: (1) Basic framework, (2) The origin of organic molecules, and (3) Organic molecules to protocells.

(1) Basic Framework[4][6] (Oparin-Haldane model):

This involves the following properties and stages of evolution:

- A reducing atmosphere of hydrogen, ammonia, methane, water, carbon dioxide, carbon monoxide, hydrogen sulphide, phosphates oxygen and ozone.

- Electrical energy to catalyze monomers (Miller-Urey expt.)
- Phospholipids --->lipid bilayers (component of cell membranes.)
- Polymerization of nucleotides to form RNA (nucleic acid first).
- First Ribosome created ---> better Protein synthesis.
- Proteins synthesized to become first biopolymer.
- The synthesis of protocells from basic components is hypothetical and has not yet been achieved, but 3 stages have been proposed:

1 Formation of Monomers.	) In-vitro evolution of
2 Monomers —> Polymers[4]	) pre-biological polymers has
	) been demonstrated.
3 Molecules —> Cells.	

(2) The Origin of Organic Molecules[4][6][7]:

Terrestrial origin - By synthesis from basic elements, through the effects of energy - singly or in combination (7.10) - on a reducing atmosphere (7.9), as in the Miller-Urey experiment (8.4).

Extraterrestrial origin - Panspermia (chapter 9)

An important assumption in Oparin's 'soup' theory, was that in an initially sterile environment, the accumulation of organic molecules would provide favorable conditions for evolution, though just how monomers polymerize to polymers is not known.

Radioactivity is a possible energy source, for example from tidal concentrations of uranium at high-water marks, on primordial beeches.

Some interesting work in the 1980s, however, has shown that external energy (as in the Miller-Urey experiment), may not always be necessary. The innate chemical energy of certain minerals, for example, could be used to synthesize organic molecules and polymers, to form self-replicating entities, which would predate the life forms known to exist today.

Replication of DNA and RNA is temperature dependent, and could have taken place without the need for enzymes under thermal equilibrium conditions on ocean surfaces, while self-assembly can occur spontaneously in RNA molecules, due to physical factors in hydrothermal vents.

Self-assembly is a standard feature of Viruses, and it could be that life itself started out as self-assembling organic molecules.

(3) Formation of Protocells from Organic Molecules[6][7]:

This is still one of life's great unanswered questions - which came first, 'genes' or 'metabolism'? There are many hypotheses, and hybrid models are now coming into favour:

One of the earliest ideas, before the discovery of genes and nucleic acids, was suggested by Oparin himself, in 1924, who proposed a form of primitive self-replicating vesicles. Several variants of this have been produced more recently, including an obscure mathematical model for the plausibility of metabolism without genes by the theoretical physicist Freeman Dyson.

Deep sea Hydrothermal Vents postulate that life began in the hydrogen-rich alkaline submarine fluids of the sea floor, where energy comes mainly from interchange of electrons between hydrogen molecules and carbon dioxide. (e.g. Extremophiles)

Geothermal activity would also provide an opportunity for the origin of life in open lakes, where there is a buildup of minerals, and a recent study comparing the spectral characteristics of sea and hot mineral water, demonstrated that life may have predominantly originated in the later, where the optimum chemical balance was found to be present.

The RNA world hypothesis proposes that self-replicating RNA molecules were precursors to life as it exists today, which is based on DNA, RNA and proteins.

Factors supporting this proposal include it's ability both to store and to catalyze information in chemical reactions (as a ribosome); its important role in processing of genetic information (in the form of DNA); and the ease of synthesizing it under conditions approximating to those of the early Earth.

RNA molecules have also been artificially produced in the laboratory, which are capable of replication, as well as providing their own template upon which copying can occur.

In addition, it has been shown that certain RNA molecules can unite smaller RNA sequences, in a form of self-replication, and if these conditions were present, then natural selection would favour the proliferation of self-catalyzing structures. Further structures could then be added later, and indeed such an RNA enzyme, capable of self-sustained replication, has been identified.

Competitive success among different replicators would have depended on the relative values of their adaptive capacities, and 'life' could be considered to have emerged when RNA chains began to express the basic conditions necessary for Darwinian natural selection to take over.

Metabolism first models do not designate replication as the essential characteristic of RNA, but argue that a stage of metabolism must come first to establish the environmental conditions that would allow that to take place.

Aerobic energy, for example, could build up ingredients initially, rather than produce genes, and as the concentration increased, each step would spontaneously convert into the next one; until eventually 'life' was established, and would take over.

Another suggestion is that atoms in a suitable environment, but driven initially by chemical energy, may gradually restructure themselves (to dissipate energy) and eventually acquiring the properties of life.

First life needs energy to bring about the reactions that form the peptide bonds of proteins, and the bonds of RNA, and there were many potential sources available as the primeval earth was forming (7.10).

Thermosynthesis, for example, was ubiquitous, where thermal energy from the suspension of protocells in a convection current (e.g in volcanic hot springs) would account for the self-organizing structure common to all origin of life models.

Bubbles of sea water provided another environment in which early evolutionary changes could take place. Wind and waves would drive debris and organic particles to accumulate in shallow coastal waters, where waves breaking along the shoreline form a foam of bubbles.

Bubbles composed mainly of water burst and dissipate quickly, but those containing amphiphiles† are much more stable, allowing time for reactions to occur within them. Some may even form a type of pseudo-membrane, and if a protein increases the integrity of the bubble, that bubble would be treated preferentially under a primitive form of 'selection', allowing those with 'best content' to rupture first; this optimizes the way 'contents' are released, which will then develop, through non-nucleated cells, nucleated cells, and eventually multicellular organisms.

A similar alternative proposes bubbles being formed in porous pumice stone (pumice rafts). These would provide a much more protected and stable environment than water bubbles, and therefore more time for

---

†   Chemical compound possessing both hydrophilic (water-loving,) and lipophilic (fat-loving) properties.

crucial reaction to take place. A further advantage would be the potential for a raft (unlike bubbles) to settle over an area of geothermal outgassing, which could greatly accelerated all of the reactions. Paleontologists have discovered evidence of pumice rafts associated with Archean life (i.e. 2,500 million years old) in some regions of Australia.

8.6 Alternative Theories for the Origin of Life[8]:

The theories outlined above are not easy to integrate into any sort of 'unified' explanation; indeed they have (aptly) been described by one biologist as 'disjointed'!

Nevertheless, the list of supplementary theories, quite unrelated to any of the above (conventional) interpretations, is long and varied - all serious proposals, scrutinized, tested in the laboratory, and published in top journals, and the following are some of the commoner examples:

1) Extraterrestrial Life: This proposes that life originated somewhere else in the universe, and that primitive forms were 'seeded' to earth at a later date (Panspermia - chapter 9). It does not explain how life began, but merely shifts the emphasis elsewhere.

2) Extraterrestrial Organic Molecules: This proposes that amino acids were formed elsewhere, and transferred to earth by comets or meteorites

3) PAH Molecules in Space: Polycyclic aromatic hydrocarbons are the most complex molecules found in space[9], and under certain conditions have been shown to transform into more complex organic compounds.

4) Subterranean life: This proposes that life originated deep within the surface of the earth's, rather than on the exposed surface itself, and today microbes certainly manage to exist and survive at depths of up to 3 miles below the surface.

5) Silicon Crystals (Clay Hypothesis): This proposes that complex organic molecules can form from pre-existing inorganic silicon crystals, but it was later shown that these would not be able to process information. (An essential requirement for life).

6) Self-Catalysts: This theory proposed that life arose from simple molecules able to act as their own catalysts (autocatalysis) and would therefore have the potential for self-replication.

7) Lipid Life: This theory postulates that the first self-replicating objects were 'lipid like', and able to store information, which would later evolve to form RNA and DNA.

8) Multiple Origins: Different forms of life could have originating in different ways, or at different times in the earth's history. Extremophiles, might be an example, having evolved under such extreme conditions, that until recently we were unaware of them.

9) God: In this context, it helps if we dissociate God from religion, and perhaps think more in terms of 'higher Intelligence' (3.8); many scientists are quite happy to compare their own field of work objectively with such a being, but shy away if Jesus Christ, or Biblical issues are raised.

If we look at the universe as a whole, even the most hardened sceptic would not deny that it's existence depends on a large number of extremely unlikely circumstances, from the critically balanced 'flat' geometry and the fortuitous coincidence of carbon resonance, to the narrow range of many of the fundamental constants, which gave rise to the concept of a 'fine tuning' (7.3).

If all of these were nothing more than pure coincidence, then the odds against such a universe existing at all must be so astronomical, that many would prefer the more logical alternative, of a creator (God or some other form of 'higher intelligence').

References (Chapter 8):

1. A. I. Oparin, The Origin of Life (Courier Dover Publications, 20 February, 2003).
2. Charles Darwin, 1 February 1871 (Quoted in reference 6).
3. A. Oparin, The Origin of Life. (New York: Dover, 1952).
4. M. Bodin, Planets and Life: The Prebiotic Era - Monomers to Polymers (JBIS, 31(4), April 1978: 129.
5. Wikipedia article (13 September 2013), Miller-Urey experiment
6. Wikipedia article (1 July 2014), Abiogenesis.
7. P. M. Molton, Planets and Life: Polymers to Living Cells - Molecules against Entropy (JBIS 31(4), April 1978: 147.
8. Wikipedia article, Other Models of Abiogenesis (Section in Reference 6).
9. B. Carey, Life's Building Blocks 'Abundant in Space'. (Space.com, 18 October 2005).

Chapter 9

PANSPERMIA

Panspermia is the hypothesis that life exists elsewhere in the universe, surviving the inhospitable environments of outer space, and able to travel (by various means) over vast distances of space and time, until eventually making contact with other bodies, such as the primordial earth, where it would have established itself as prebiotic life, and evolved conventionally thereafter, into the many forms of life which exist today.

9.1 Background:

The concept of life having come to this planet from elsewhere, is not new. It was mentioned in the writings of Greek scholars as early as the 5th century BC, and intermittently throughout medieval times. Lacking any sort of scientific basis, however, it was never a serious proposal, until the work of pioneers such as Helmholtz and Lord Kelvin, in the 19th century, established many of the foundation principles of modern physics and chemistry today.

It was the Swedish chemist Svante Arrhenius, at the turn of the 20th century however, who first put panspermia on a serious footing, by proposing that life could have been carried as seeds or spores between planets[1], and therefore, in principle, could be widespread throughout the universe. He was a Nobel laureate, and later director of the Nobel Institute, with a wide range of interests, from auroras and zodiacal lights, to the radiation pressure of sunlight, which he later incorporated into his ideas.

It was Fred Hoyle and his former student Chandra Wickramasinghe, however, who became the strongest advocates[2], and in 1974 rationalized the whole concept of panspermia, with a number of innovative proposals: that meteorites and comets, for example, could act as vehicles for the transport of 'seeds', whether living or fossilized; that in certain circumstances, dormant viruses or even DNA and RNA molecules, could survive the hostile conditions of outer space, unprotected; and they were the first to propose that interstellar dust was largely organic in nature - which has since been amply verified.

Their most controversial proposal, however, was that over time, the accumulation of viruses and bacteria entering the earth's atmosphere could have been responsible for major disease epidemics over the years, including for example, the 1918 'flu epidemic, AIDS and mad cow disease This suggestion is no longer generally accepted, but it has certainly not been disproved, and many prefer to withhold judgement for the time being[2]

.Hoyle was a unique individual in many way, aside from intellectual, and also a lifelong confirmed atheist. Yet it is widely believed that two things caused him to change his beliefs - the extremely improbable existence of carbon resonance(7.7), without which organic life would not have been possible; and the complexity of biological life forms:

> "....for if one proceeds directly, without fear of incurring the wrath of scientific opinion, one arrives at the conclusion that biomaterials, with it's amazing order, must be the outcome of intelligent design.........for I can think of no other possibility"[3].

Holye was also fond of melodramatic analogies to prove his points. For example, that the chances of obtaining the required set of enzymes for even the simplest living cell, without panspermia, was infinitesimal compared to the number of atoms in the known universe, and therefore that:

> "The notion that not only the biopolymer but the operating program of a living cell, could be arrived at by chance in a primordial organic soup here on the Earth, is evidently nonsense of a high order......"

....and went on to add that:

> "The suggestion of a 'guiding hand' left him "greatly shaken"

Among other favorites, were comparing the likelihood that even the simplest cell could have arisen spontaneously, without panspermia, to the chances "of a tornado sweeping through a junk-yard and assembling a Boeing 747 from the rubbish therein"; or, comparing the chances of obtaining a single functioning protein, by chance combination of amino acids, with those of a solar system full of blind men, all solving Rubik's Cube simultaneously[4].

Chandra Wickramasinghe remains a firm supporter of panspermia, and in particular the possibility of diseases arising in this way. In 2003, together with colleagues, he published a letter in The Lance proposing a possible extraterrestrial origin for the virus that causes severe acute respiratory syndrome (SARS)[5], rather than the generally accepted view that it came from chickens.

The basis for this idea, however, has been strongly challenged, and today there is really no positive evidence to correlate any known diseases with an origin of this sort.

Nevertheless, the concept of 'main-stream' panspermia has a firm basis of plausibility, together with a considerable amount of indirect evidence to support many of it's underlying assumptions, both with respect to potential methods of 'transport', and to the 'survivability' of micro-organisms, under the most extreme conditions possible[6],.

Even the physicist Stephen Hawking, expressed support for panspermia, when he stated that:

> "Life could spread from planet to planet or from stellar system to stellar system, carried on meteors"[7].

9.2 Implications:

Although interplanetary 'movement' is the more conventional context in which panspermia arose, that would be a very restricted interpretation when we come to consider the question of how life may have originated in the first place. Indeed, from the past evolutionary history of the solar system, and the knowledge we have today of all that biogenesis must have involved, we can be fairly sure that none of the other planets in the solar system would have been suitable for that purpose.

Nevertheless, the position was very different in 1903, when Arrhenius introduced his ideas, and Percival Lowell's theories of 'canals' and intelligent life on the planet Mars, were very much in vogue[8] (chapter 12). However, at the time, these ideas were interpreted not so much in the context of 'source', but rather as an indication that life appeared to be common, and therefore would almost certainly exist elsewhere in the universe.

It was also a convenient background for Arrhenius to present his hypothesis, with respect to the implications of panspermia, but without becoming involved in the question of definitive origin; indeed, he made it clear from the outset that panspermia concerned only the 'presence' of life on earth, and not it's 'source'- either with respect to 'where' or 'how' that might have come about.

It was a few decades before the ideas of Percival Lowell were finally laid to rest, but from the 1920s onward, as instruments and technology improved, it became clear that with the possible exception of micro-organisms, none of the other planets could have harbored any more advanced forms of life.

As an interesting corollary, however, it has been pointed out that if ever we do manage to confirm something analogous to viruses or microbes on one of the other planets (or their satellites) then before accepting this as evidence of extraterrestrial life, we must first exclude the possibility of 'reverse panspermia' - i.e. that such life forms may have come from this planet in the first place, rather than the other way round. Barring the possibility of totally unknown organisms, therefore, which could never to have existed on earth, this might prove very difficult to do.

The era of terrestrial volcanic activity violent enough to bring that about has long since passed, but we now know a great deal more about the life history of meteorites which have landed on earth from other sources in recent times, and would have been very suitable for the transport of micro-organisms. Many of these have been orbiting the sun for billions of years, or even possibly residual from the prot-planetary disc out of which the planets initially formed, before finally being gravitationally captured, and drawn in to impact on the surface of some planet or satellite.

Hence, the fact that the earth's surface has been relatively quiescent for so long, does not preclude the possibility of volcanic activity in the distant past, which could have ejected 'contaminated' material into space.

9.3 Radiopanspermia:

This was the form in which Arrhenius first proposed his original hypothesis, using the pressure of sunlight as a means of propulsion. This was not only a convenient natural source of energy, but he had also calculated that provided a particle was smaller than about 1.5 ìm, it could propagate rapidly under the influence of radiation pressure from the solar wind. This seemed ideal for his purpose, since most bacteria, for example, are barely one third of that size.

Even so, its effectiveness falls off quite quickly with increasing particle size, and in practice it would probably have been limited to occasional spores. A more serious objection was the damaging effects of UV radiation and X-rays, which inevitable accompany sunlight exposure..

Irrespective of these objections, however, Wickramasinghe has always been an advocate of transporting individual or small clumps of bacteria, and believes that overall, far greater numbers of microbes will eventually be transferred in this way, than by other means, such as meteorites.

Nevertheless, recent experiments to test out the effects of exposure to harmful radiation, have demonstrated that isolated spores are killed off within seconds of unprotected exposure, while the same spores, if properly shielded from UV radiation, can survive in the space environment for up to 6 years.

Radiopanspermia may take place occasionally, but it is no longer considered important in relation to alternative mainstream proposals.

9.4 Transport:

Meteorites are now regarded as the most important method for transferring organic molecules, and possibly primitive micro- organisms, both on theoretical grounds, and as the only objects from space which we can actually handle and examine.

A great deal of information, for example, has been gained from radioactive dating, structure formation, chemical and isotopic composition, and electron microscopy - for example in the Allan Hills Martian meteorite, which was thought to show possible fossilized micro-organisms, as the first conclusive evidence for the existence of extraterrestrial life. It is also one of the few meteorites known to have come from a 'wet' era, when water (essential for any form of life) was present on the surface of Mars. It generated a great deal of interest at the time, but has since been seriously questioned on the grounds that morphology alone is notoriously subjective, and has led to numerous errors of interpretation in the past[9].

Meteorites are extremely common, and though their overall distribution will be quite random, they are more easily collectable in some areas than others - for example, the deserts in central Oman, where over a recent ten year period, up to 5,000 specimens were obtained, including large numbers of lunar and Martian meteorites[10].

Such bombardment is as old as the solar system itself, and we know from direct observation that every visible planetary surface is peppered with impact craters - including those of Mercury, Mars, the moon, and larger satellites of the gas giants.

Not all meteorites have a straightforward history, however, and some may even have been ejected into space more than once, before finally arriving on earth. The Allan Hills meteorite, for example, is thought to have been broken up initially on the surface of Mars by other meteorites, about 4 billion years ago.

It was then blasted into solar orbit from the Martian surface only 15 million years ago, and impacted on the earth about 13,000 years ago. These dates were established by a variety of specialized radiometric dating techniques.

Venus is very different from Mars. It has an extremely thick dense atmosphere, mainly consisting of CO_2 and nitrogen, and a volcanically active isothermal surface at many times the temperature of boiling water. This may have been different in the distant past, however, and possibly more resembling that of the earth, together with significant amounts of liquid water.

Nevertheless, although over the years Venus would have been subjected to much the same meteorite bombardment as the earth, only a minority of these would actually penetrate to reach the surface, and probably few organisms would have been able to survive. However, from what we are now discovering about extremophiles (8.5,6) on our own planet, we can never exclude the possibility that similar organisms might have been present, able to survive and adapt, even to the hostile conditions on the surface of Venus.

Comets (5.9) are also now widely accepted as vehicles for the potential transport of micro-organisms. Many of these are thought to have originated in the Oort cloud (5.4), and consist of little more than a rocky core, or nucleus, perhaps up to 20 miles across, together with dust, water, ice and frozen gases - popularly described as 'dirty snowballs'.

As they approach the vicinity of the sun, volatile material begins to vaporize, and streams out behind the comet, due to radiation pressure of the solar wind, to form the characteristic tails, which have been know to extend for up to two to three astronomical units (over 300 million miles).

It is thought that comet bombardment of the primordial earth may have been responsible for the vast quantities of water that now form the oceans, and in the process, organic material and micro- organisms, could have been assimilated in to the pre-biotic soup.

Significant quantities of organic molecules have now been detected in comet tails, and it is possible that amino acids could form on the surface of the core of a comet - perhaps even with the potentials for 'shock synthesis' of proteins at a later stage.

The amino acid glycine has been found both in comets and in meteorites, and a very similar molecule has also been detected in gas clouds near the galactic center.

All of these features support the concept of panspermia in general, and the active role of comets and meteorites, as 'vehicles' of the transport for organic material, in particular.

Other mechanisms which could bring about or influence the spread of primitive life forms and organic material, might include smaller space debris and meteors, gravitational fields, magnetic fields, or even contaminated space probes.

Rigorous policies are now in force for planetary protection, including dry heat sterilization, but none of these can ever be foolproof, and as soft landings extend, to include other planets and satellites, the risk of 'reverse panspermia' increases too. Unlike natural 'reverse panspermia', however, which would have taken place in the distant past, and therefore might be wrongly interpreted as extraterrestrial life today, 'probe' borne organisms could not be confused in this way at the time of transport; but they could be misinterpreted by any subsequent mission.

9.5 Environment:

Aside from the method of transport, we also have to consider the environment in which that takes place, For planet to planet panspermia therefore (lithopanspermia), there are three stages involved:

(1) Initial injection into space. Normally the result of volcanic activity, this would involve both impact and acceleration forces.

As noted earlier, Martian meteorites are common, and these would have been subjected to high initial 'g' forces, to achieve the Martian escape velocity of almost 3mps, as well as thermal and possibly friction effects, as Mars may have had a significant atmosphere in the distant past, when these events would have taken place.

(2) Transit. Panspermia was initially visualized on a planet to planet basis, within the solar system; however, there is no reason in principle why it should not apply more widely, and hence we now have to consider not only the interplanetary environments, but also interstellar and intergalactic space as well.

By terrestrial standards, all of these approximate to perfect vacuums, with only a very tenuous content of atoms, ions and radiation, which we discussed earlier in chapter 2.7.

Within the solar system, however, radiation (UV and X-rays) are the main hazzard of space exposure, together with high energy charged particles from the solar wind (especially associated with solar flares and coronal mass ejections) and cosmic rays.

The effects of these will depend on a variety of factors - shielding (within the substance of the meteorite), physical condition of micro-organisms (seeds, spores, and wether live, dormant or fossilized), while isolated or exposed molecules (e.g. DNA, RNA), are at greatest risk. For photosynthetic organisms, shielding is a balance between 'protection' and 'exposure', to ensure adequate sunlight energy during transport, while orientation of the surface of the meteorite to the solar vector, would be important for the same reason.

(3). Arrival (atmospheric entry). Once 'captured', a meteorite accelerates inwards, under the gravitational influence of that planet, to enter the tenuous upper layers of atmosphere at very high speeds, perhaps of the order of 12 -15 mps, for an earth sized planet. As it descends into thicker layers, deceleration forces and friction rapidly increase, temperature escalates, and the meteorite may start to vaporize or disintegrate.

Large acceleration and jerk exposures will be common both on exit and entry, and have been estimated to be of the order of 6x109 m/s2 for a planet such as Mars.

Much more attention is given in the literature to 'exit' stresses (escape velocity) as opposed to those on 'reentry', but nevertheless, the impression that organisms able to survive the former will automatically survive the latter, where very high temperatures may be involved, is questionable, e.g:

> 'The chief danger to potential panspermic life is not vacuum, acceleration or even cosmic rays.....but that life can survive the extreme temperatures of re-entry'[11].

This is an unsourced statement, and many might disagree with it, but it does draw attention to an important issue with respect to practicality.

The net consequences of all of these factors determines how much of the meteorite will eventually reach the surface, while how much organic content survives, will depend on how effectively it is protected and shielded, both from the physical effects of transit, and the impact shock when it finally comes to rest. Live organisms, as most vulnerable, have the worst prognosis.

If organic content was distributed throughout the body of the meteorite, the probability is that some may survive. Thereafter, long term future will depend on the local environment, and atmospheric composition in particular (which was strongly reducing in the primordial earth). There is also the possibility that other life forms may already be present, in which case the next challenge could be a Darwinian struggle, as to which is best fitted to survive and evolve.

As it is, life which exists on earth today, could just as likely be the consequence of such a sequence of events, as it's potential instigator.

9.6 Cosmic Panspermia:

This covers panspermia outside and beyond of the solar system, and theoretically falls into two categories.

(1) Interstellar Panspermia.

At first sight this may seem a very different proposition from lithopanspermia, and the relatively confined environment of the solar system. However, that would be a misunderstanding of the importance of size. The 'initial' and 'terminal' events will still be essentially the same as outlined above, and although the physical separation between 'start' and 'finish' is immeasurably greater, it is 'time' rather than 'distance' which really matters.

We have no direct way of comparing the two, but it could be that an interstellar journeys in fact may be no longer than the time spent by known meteorites, endlessly orbiting the sun, perhaps even for the lifetime of the solar system.

There is one prediction we can make, however. Meteorites will always be subject to random gravitational influences, and these could shorten or terminate there existence in space at almost any time - the Allan Hills meteorite, for example, impacted earth after a mere 15 million years.

For interstellar journeys, by comparison, once started, there can be no diversions or turning back. At an escape velocity of perhaps 5 -10 mps, it could take many billions of years just to reach the vicinity of the nearest star, and even then, if we compare the density of stars in the vastness of cosmic space, with that of planets within the solar system, a given journey might progress almost indefinitely before reaching another 'body'. However, it must always do so eventually, because no line of sight trajectories through any part of the universe, can ever be 'open ended'.

On average, therefore, if we assume that 'journey times' could be roughly the same for 'stellar' as 'solar' panspermia, it is the 'en route' circumstance which are important, and within the solar system, there is a greater variety of risk factors - changing solar activity, directional, radiation, UV etc, all of which are randomly variable. Nevertheless, the magnitude of risk factors in interstellar space, completely dwarfs anything which could be encountered in the solar system - very high energy cosmic rays, up to 3×10^{20} eV, or about 40 million times greater than the energy of particles accelerated by the Large Hadron Collider (i.e. comparable to the energy within milliseconds of the Big Bang itself).

The only other risk factor, but insignificant by comparison, might be depravation of light energy for photosynthetic organisms - though they would probably not survive the radiation, unless hibernating and deeply buried within the substance of the meteor.

There is one other significant difference between solar and stellar panspermia. The solar system is strategically situated in what amounts to a 'galactic habitable zone', on the rim of the Orion arm of the milky way galaxy, about 25,000 light years from the galactic center(5.3). This is not random, and among the advantages of such a position are the fact that for most of it's 250 million year cycle of galactic rotation, the solar system lies largely outside areas of high stellar density (where supernovae abound, together with gravitational instability and high levels of radiation).

Hence, there are potential advantages and disadvantages, depending on which direction an interstellar meteorite happens to be traveling. If, for example, it is heading towards the galactic center, as it leaves the 'habitable zone all risk factors increase progressively, compared with those in the opposite direction.

In reality, however, there must be a 'cutoff' point, with respect to the galactic center, within which life would not be able to exist anyway. There is no way to quantify risks of this sort, because we know too little about the precise nature and extent of galactic habitable zones, beyond the fact that they will depend both on source, and the direction in which a meteorite is heading.

Assuming successful completion of a mission, there would still be a high risk of mutations, even in organisms which might otherwise appear to be healthy, and what the effects of these might be we can only guess - possibly none at all; causing disease, as Wickramasinghe and colleagues suggests; or perhaps total incompatibility, and all that that could entail.

2) Intergalactic panspermia.

This is a theoretical possibility only. From all that we have outlined above, there is clearly no reason in principle why spores and organisms should not be transported between galaxies in essentially the same sort of ways. In practice, however, it may not be that simple, for we know nothing at all about physical contact over distances as large as those which separate an average pair of galaxies, and the reality could well be that such a journey might simply not be possible.

Hence, for any form of panspermia transport, assuming it could take place at all, at the inevitable sub-light velocities involved, the time scale for such a journey would be enormous, and perhaps even approaching the lifetime of the universe itself†.

It does, however, raise some general questions about life in the universe. Starting with the certainty of our own existence, it is not difficult to rationalize how life could spread out within the Milky Way. If we accept too that physical communication between galaxies is probably not possible, we are still faced with the statistical probability, given all the recent findings of the Kepler space observatory, that ours is not the only inhabited galaxy.

Hence, there are two possibilities: either life had a single origin, at one unique point in space and time, from which it evolved, and eventually spread to colonize the whole universe - in which case, why do we not know about it, and how did it travel between galaxies? Or alternatively, life was 'multi-center' with at least one focus of origin in every one of the billions of galaxies known to exist - which is difficult to reconcile either with random chance or rational choice.

To do something a billion times, just to achieve uniformity, when one single event, able to spread, repeat and proliferate could eventually achieve the same objective, seems in complete conflict with the possibility of life being the work of an 'intelligent entity'.

That would be an anthropomorphic stance, nonetheless, with the implicit assumption that all forms of life must necessarily be similar to our own.

But what if other types of life exit elsewhere in the universe, that we know nothing about - perhaps because it was never intended they should mix in the first place. The universe might, for example, be a 'balanced' collection of different species, with intergalactic distances so large, just to ensure it remains that way. And depending on how important segregation actually was, there could be other barriers in place as well, not only to prevent spread, but perhaps even to preclude any form of contact,

---

†   The equations of special relativity preclude travel at speeds greater that of light. This has been confirmed in many different ways, and no exceptions have ever been found. Theoretical proposals that the speed of light may have been different in the early univeres[12] have not been confirmed. NASA has been studying the possibilities of FTL travel since 1995, but so far without success[13].

including communication - which might be one explanation for Fermi's paradox.

That would be a 'designer universe' concept, and while it might be difficult for human beings to understand alien logic, uniformity would surely have been a great deal simpler. Hence, given the universality of 'carbon life potentials' (together with 'the geometry of old age') it would be surprising if the 'end result' (organic life) was not equally common and equally widespread.

However, there is a third possibility - that life is not common after all, and that planet earth is indeed the only place in the whole universe where it exists..

9.7 Directed Panspermia:

This is pure science fiction, where it belongs, and we include it here only to avoid confusion.

It proposes the deliberate and targeted spread of life by an advanced extraterrestrial civilization, either to start life on earth, or as a means by which we ourselves could seed life onto new planetary systems - perhaps from among the many potentially suitable candidates recently identified by the Kepler space observatory.

The concept owes it's origin largely Carl Sagan in 1966, and not as sometimes stated, to the Nobel laureate, Francis Crick, who made a similar proposal in 1979.

A very detailed description, including design, strategies and targets, propulsion systems, etc - everything in fact required for a real life mission, can be found in Wikipedia[14].

9.8 Astrobiology:

This has now replaced older terminology - Space Biology, Exobiology, and Exenbiology - to include the study of all aspects of life within the universe, and hence has an extremely wide, multi- disciplinary remit. Topics embraced include physics, chemistry, astronomy, biology, molecular biology, ecology, planetary science, geography, and geology - focusing particularly on the possibilities of life on other worlds, properties and characteristics of 'alien biospheres' in so far as these might be different from our own, and risks and hazzards associated with different forms of panspermia transport.

The astrobiology journal is a peer review journal and respected authoritative outlet for work which might not readily fit with the more conventional constraint of parent disciplines.

This is undoubtedly an area with huge potentials for study and research, aptly described as at 'the forefront of plausibility', for surely it is only here that we are likely to uncover those 'missing links' necessary to extend conventional science into areas which have hitherto seemed 'out of bounds'.

It is also encouraging to note a willingness to take seriously those pillars of science fiction (Stapledon, Clarke, Asimov and Sagan) where fact and fiction, both combined by, and obscure by, originality, must await the attention of an unbiased mind, to separate the two - for there is now ample precedent that 'today's science fiction can become tomorrow's science fact'[15].

Astrobiology is now a well recognized discipline in it's own right, with the central issue of wether life exists elsewhere or not, a rational and innovative challenge.

NASA established it's Astrobiology program in 1970, incorporating both Exobiology and SETI (Search for Extra Terrestrial Intelligence) - which began 10 years earlier, as project Ozma, to scan selected radio frequencies for possible evidence of 'intelligent signals', and has been ongoing ever since.

SETI is the only way we are ever likely to obtain direct evidence of 'intelligence' elsewhere, which statistically, from present knowledge, we might expect to be common. Yet in 50 years of continuous monitoring 24/7, with one possible unconfirmed exception†, we have drawn a complete blank - a rather surprising result, which is now implicit in the Fermi paradox.

Martian life is the dominant focus of astrobiology today, with an almost limitless supply of meteorites available for study, and a total of 51 Mars missions since 1960; 31 of these were failures, and as of March 2014 there are 2 US and I Chinese probes active on the Martian surface. Manned missions will undoubtedly happen with the next few decades, and training under simulated Martian conditions has already begun.

Extensive 'laboratory' studies aim to simulate every possible type of environmental situation, and have been carried out using decompression

---

† The 'Wow! Signal', 15th August 1977.

chambers, centrifuges, zero-g exposure, diving bell and deep sea underwater studies, and a great deal of custom built equipment to simulate the many different aspects of radiation exposure (UV, X-rays), high energy particles, thermal, pressure and gas composition, among others.

All of these studies have involved both human and animal exposure, together with many different forms of micro-organisms, bacteria, viruses, seed and spores - live and dormant.

Time is an important aspect of panspermia, with exposures over billions of years, which can never be duplicated, but comparative studies of 'exposed' versus 'shielded' conditions can be scaled down and adjusted, and we now have a huge amount of information regarding time dependent variables, from which to extrapolate.

Through studying meteorites, and recent data available from Martian surface probes, we probably now know more about Mars - climate, geology and evolutionary past - from a distance of 45 million miles, than we do about the sea floor of some of our deepest oceans, a mere 5 or 6 miles below the surface!

Direct exploration will now be central to how we study and explore the solar system in future, though always constrained by the uncertainties of government funding, while conventional astronomy (together with advancing skills and technology) will otherwise remain our only access to the universe beyond.

Nevertheless, this has been our sole source of information up to now, with spectroscopy, for example, responsible for the huge amount of knowledge we now have about the chemistry and biology of the universe at large.

There can be little doubt that astrobiology and physical sciences will progressively merge and eventually overlap, which may simplify funding (to the benefit of all), and help to focus combined attention on whichever single issues matters most.

9.9 Evidence:

Both astrobiology and SETI have a common objective - just one single piece of incontrovertible evidence that extraterrestrial life exists. And in so far as they represent entirely different and unrelated approaches - from the 'microbiology of origins' at one end, to the 'sophistication of advanced intelligence' at the other - the chances of success are increased.

Neither, however have so far been successful. SETI, with a more limited remit, has devoted the most time (24/7 over 50 years), and so far drawn a complete blank; while biologists, with a broader remit, and variety of approaches, have officially done no better.

There are a large number of interesting specimens available for study, but the Allan Hills meteorite (9.4) is the only one where a serious element of doubt still exists. Unfortunately, we do not yet have the technology which might settle the question, and so long as interpretation remains subjective, there will always be genuine differences of opinion.

Some of the other examples[16], topical at the present time, and which might be regarded as 'undecided', include the following:

- In 2001, two researchers from the university of Naples were able to revive a micro-organism wedged within the substance of a very ancient meteorite, which survived sterilization by several means, and had DNA 'unlike any on earth'. Based on genetics and morphology, they presented a number of papers, claiming an extraterrestrial origin, but there is no record of any follow up. As a general rule, however, 'compelling evidence' which is not taken further (very common), is usually an indication of 'second thoughts' and genuine doubt.

- In 2001, the Indian Space Research Organization obtained living cells from air samples at an altitude of 25 miles, where terrestrial life would not normally be found. NASA scientists however, disputed this latter statement, and subsequently the Indian Astrobiology Center "found no evidence of distinct adaptations expected in micro-organisms occupying a cometary niche"

- In 2007, a camera brought back by Apollo 12, from the lunar lander probe Surveyor 3, was found to contain a live streptococcus bacterium. Although subsequently there were strong reasons to suspect contamination, NASA refused to accept them, and sticks with it's original report of 'micro-organisms on the lunar surface'.. However, this too does not seem to have been taken further.

- In 2013, Chandra Wickramasinghe and a number of colleagues, reported finding shapes resembling fossil diatom frustules in a carbonaceous meteorite. Although the nature of this specimen was subsequently questioned, his team was able to produce very strong evidence of authenticity. There has been a good deal of further research related to this claim, together with several papers, but again no evidence of what the final outcome may have been.

In the opinion of one Wikipedia editor, whose job it is to evaluate objectively the most probable interpretation for borderline claims and topics, panspermia is a serious proposition, well supported on theoretical ground, and not in conflict with accepted science, though without one piece of proven supporting evidence.

On that basis, he suggests that panspermia should no longer be regarded as a 'fringe' topic (which is more often than not the case), but rather deserves the same prominence and attention as any other serious and plausible proposals of this sort[17]. We strongly agree with that.

9.10 Research:

The question of whether microorganisms, such as bacteria and viruses, or even spores and seeds, can survive in the hostile conditions of outer space, has interested biologists since spaceflight became a reality in the 1960s.

Today, such questions are no longer hypothetical, but real practical issues, with implications for spread and contamination, panspermia (both ways), human contact, medical, sterilization of space craft, and the whole question of adaptation and survival under extremes of environmental exposure, both short and long term..

The earliest practical studies involved the exposure of bacteriophages and Penicillium spores, to the space environment, during the Gemini missions in 1966, while about the same time, both the Soviets and the US began similar studies in low earth orbits.

Interest in panspermia was very high at that time, and an important motivation to early studies, where survival under space conditions might indicate whether or not it would be possible in practice. However, as we have seen, there is a great deal more to survival under hostile conditions, than simply 'exposure'.

Today there are a large number of active projects, both terrestrial and in space, but all essentially concerned with the same thing - tolerance of life to environmental extremes.

It is only within the past few decades that we have come to appreciate that there are regions below the earth's surface, where environmental conditions are at least as harsh and hostile as many which exist in space, and even more important, there are life forms (extremophiles) which exist and survive naturally under these conditions, their physiology and metabolism refined and adapted to individual circumstances, and entirely different from those of surface life.

Biologists are as keen to study (and learn from) these, as their 'space' counterparts are to the study survival and tolerance of 'normal' life forms under extreme conditions. On earth, these environments are all below the surface, but differ widely from each other, depending on the region concerned. By there very nature, however, many are almost as inaccessible as space itself. Hydrothermal vents, for example, where extreme conditions are found, exist in the deepest oceans, up to 3 miles down and pressures of over 300 atmospheres. No human could survive these, and exploration is with deep sea submersibles, such as Alvin.

By contrast, organisms have been found in warm rocks, deep below the earth's crust, and frozen into blocks of ice, 2 miles below the surface of lake Vostok in Antarctic.

Organisms have also now been found which are able to thrive and survive in the depths of the oceans, pressures up to hundreds of atmospheres, ice, boiling water, acid, the water core of nuclear reactors, salt crystals, toxic waste and in a range of other extreme habitats. Surprisingly, there is no record in these studies of an environment in which no form of live could survive, and if we are to take that at face value, it gives encouraging support to proponents of panspermia.

In many respects, space studies are easier to carry out, and a variety of projects, recent or ongoing at the present time, include:

- BIOPAN, a Russian experimental facility to study the effect of the space environment on biological material after exposures of between 13 to 17 days. Six missions have been flown so far, microbes, bacteria, lichens and even one micro-animal (tardigrades). Results in general confirmed other findings, of an overall high survival to a wide range of harsh environments, including cosmic rays,
- EXPOSE is a multi-user facility attached to the International Space Station, to study medium term exposures to the outside

environment, including microbes, which might have practical uses for life support, so far up to 553 days.

- The Rosetta spacecraft is due to place a small lander on the core of a comet, later in 2014, and then to monitor that over the next two years, particularly for evidence of dormant microbes, as hypothesized by panspermia.

- A specific project to test out lithopanspermia, by exposing microbes in a capsule attached to a Russian spacecraft, on a 3 year interplanetary round trip, unfortunately suffered technical difficulties and the launch was unsuccessful.

9.11 Conclusions:

Panspermia is not an 'origin of life' theory, and it's occasional comparison with abiogenesis is misplaced.

It is a proposal that life exists elsewhere within the universe, and through the dissemination of seeds or spores, might have been responsible for it's presence here on earth.

It is also consistent with the possibility that life arose first on the earth itself, and subsequently could have spread outward from there, to colonize other parts of the universe.

Abiogenesis and pamspermia are not mutually exclusive, especially if both life forms were organic - as would seem most probable..

Confirmation of extraterrestrial life would neither support or preclude panspermia.

References (Chapter 9):

1. S. Arrhenius, Worlds in the Making: The Evolution of the Universe. (New York, Harper & Row, 1908).
2. F. Hoyle, C Wickramasinghe, Evolution from Space (Cambridge University Press, 1982).
3. F. Hoyle, Evolution from Space: A Theory of Cosmic Creationism (Omni Lecture, Royal Institution, London, 12 January 1982).
4. F. Hoyle, The Intelligent Universe (Dorling Kindersley, 1983).
5. C. Wickramasinghe, M. Wainwright, J Narlikar, SARS - a clue to its origins? (Lancet, May 24, 2003).
6. L. J. Rosthschild, R. L. Manuinelli, Life in Extreme Environments (Nature, 2001,409, 1092-1101).
7. S. Hawking, 'Origins' Symposium (2009).
8. P. Chambers, Life on Mars - the Complete Story (London, Blandford 1999).
9. J. Manuel, G. Ruiz, Morphological Behavior of Inorganic Precipitations Systems (SPIE.Digital Library, 30 December 1999).
10. Wikipedia article (10 September 2003), Meteorites.
11. Wikipedia article (6 September 2013), Impact Survival ('Talk' page).
12. J. Magueijo, Faster than the Speed of Light (Heinemann, 2003).
13. I. A. Crawford, Some Thoughts on the Implications of Faster-than-Light Interstellar Space Travel (Quarterly Journal of the Royal Astronomical Society 36 (3), Sept 1995).
14. Wikipedia article (3 September 2013), Directed Panspermia.
15. L.M. Kruss, The Physics of Star Trek (Harper perennial, 1995).
16. Wikipedia article, (11 September 2013), Panspermia (case studies).
17. Wikipedia article (11 September 2013), Panspermia ('Talk' page).

Chapter 10

ANTHROPIC PRINCIPLE†

In it's most general form, this is a philosophical principle which seeks to correlate life with those physical properties which are essential for it's existence (the fine tuned universe, 7.3) yet defy all odds of probability in order to exist themselves.

It is not an easy concept to grasp. Life is seen as a conscious intelligent entity, able to observer the nature of the universe, which in turn must be compatible with it, and the many ways in which this relationship can be expressed reflect the principles of anthropic philosophy.

10.1 Background:

It was the evolution of intelligence, which first led man to question his origins and background, the nature of existence, and the importance and significance of life itself. His privileged position, with respect to other life forms, seemed incongruous compared with the insignificance of

---

† 'Anthropos', Greek for 'human'

earth, and his continued failure to find evidence of life elsewhere, served only to reinforce this anomaly.

It was the biologist Alfred Russel Wallace, however, in 1904, who first suggested that the complex nature of the universe looked very much as though it was somehow intended "for the orderly development of life", and essentially introduced the concept of a Fine Tuned Universe.

In 1913, the chemist Lawrence Henderson also drew attention to this in his book The Fitness of the Environment[1], which pointed out that life depended entirely on very specific environmental conditions, while the American astrophysicist Robert Dicke, in 1957, refined that even further, when he noted that if the values of the fundamental constants had been different by even the smallest amount, life would not have been possible, and there would have been no human beings in the universe to consider the problem.

It was against that background that the anthropic principle was first proposed in 1973, by an astrophysicist, Brandon Carter, who rationalized the improbable physical conditions of the fine tuned universe with one sweeping generalization - that unless they were exactly as they are, it would not have been possible for life to exist.

'Fine tuning' was seen as 'selection bias', so that life could only appear in a universe capable of supporting life. In other words, the physical nature of the universe 'had to be compatible with the conditions necessary for life, or it would not be have been able to appear in the first place'.

This is the sort of simplistic logic which permeates the whole of the anthropic principle, which comes in many different forms, can be interpreted in many different ways, and in consequence of that, has been used to justify some of the deeper mysteries of existence and creation.

Nevertheless, as originally proposed, it is little more than a truism, and difficult to see how that has come to be so widely interpreted in ever more complex ways, to its present revered position as a 'cosmic philosophy', giving insight and meaning into hitherto unexplained aspects of life and the universe.

Of the many different version which now exist, tautologies are common, with 'conflict and contradiction' resolved only by the individuality of interpretation, and application to such diverse topics as panspermia, Fermi's paradox and probability theory.

Yet underpinning all of these, the self-evident truth, that:

'things are true because they are true, and had they been otherwise they would not have been true'

which irrespective of semantics, virtually paraphrases the 'anthropic wording' in which some versions are expressed.

'Interpretation' in fact lies at the heart of much of anthropic reasoning, especially where vague or ambiguous concepts are involved, and accounts for much of the variety now inherent in the many different versions of Carter's original concept, which started out more as simple observation, than a formal proposal.

10.2 Format and Variants[2] [3]:

There were two versions of the Anthropic Principle as originally proposed by Carter:

(1) Weak Anthropic Principle (WAP):

We must take account the fact that our location in the universe is necessarily privileged, in so far as being compatible with our existence as observers.

(2) Strong Anthropic Principle (SAP):

The Universe must be such as to admit the creation of life (i.e. observers) within it at some stage.

Barrow and Tippler, in their book The Anthropic Cosmological Principle[8], proposed their own, alternative versions, of both of these, namely:

(a) Weak version (B&T):

The values of fundamental constants are not equally probable, but dependant on two conditions - that there are sights where carbon-based life can evolve; and that the universe is old enough for it to have already done so.

This differs significantly from Carter's weak principle, by including 'physical constants' (which Carter includes by inference, only in his SAP), and by limiting life to organic forms only.

(b) Strong version (B&T):

The Universe must have those properties which allow life to develop within it at some stage in its history.

This was then further interpreted to mean that 'observers are necessary for the universe to exist', and also that 'there is only one possible universe designed specifically for observers to exist'.

From these definitions, we can draw a number of conclusion:

1. The role of observers in the universe, able to observe and know of their own existence, is fundamental to all interpretations.
2. Observers are synonymous with 'life'
3. There is a purely coincidental hierarchy of relevance, which can be expressed more simply in the form:

*   Life is special because it allows us to exist as observers.
*   The universe must be such as to allow life to exist at some stage
*   Physical conditions are determined by where and when life exists.
*   Physical properties of the universe must be suitable for life to exist.

4. Both 'strong' version [(2)and(b)] look remarkably similar, though Carter's original version only stipulates that 'conditions must 'admit' the creation of life, while in Barrow and Tipler's version, the universe must have 'properties necessary' for life - i.e. that life will be inevitable at some stage. Nevertheless, Some see this distinction as artificial, essentially 'splitting hairs', while the important point is that life is imperative.

There are four other versions of the principle, which bear no resemblance to Carter's original proposal, and are even more obscure and difficult to understand:

The Modified Anthropic Principle states that the problem of 'existence' is only relevant to whatever entity is able to question it in the first place, and that before man was intelligent enough to do this, the problem simply did not exist.

The Participatory Anthropic Principle, by contrast, is a quantum mechanical interpretation, and beyond the scope of this chapter.

The Strong self-sampling assumption refers to discrete intervals known as 'observer moments', and states that each one should reason as if it were randomly selected from the 'class' of all observer-moments. Analyzing 'observer-moments' can help to avoid paradoxes, but mistakes in the choice of initial 'reference class' can lead to counter-intuitive results!

The Final Anthropic Principle, which Tipler expounds in his book The Physics of Immortality[4], removes the emphasis from 'life', which by it's nature must be transient, to 'information', and states that Intelligent information-processing must come into existence in the universe, and, once it has done so, will never die out.

10.3 Concepts and Definitions:

Any philosophical proposition depends very much on subjectivity, as to how it is expressed and understood, but no matter how fine the distinction, or the form of wording used, all versions must belong to one or other of two very distinct interpretations of Carter's original concept - initially, the so called 'weak principle',which simply pointed out the importance of 'fine tuning', as necessary to make the physical conditions of the universe compatible with the existence of life in it's present form; and a strong version, which argues that not only are these conditions necessary, but that the universe had to evolve from the outset to ensure that such conditions existed, so that life would be inevitable at some stage in it's future.

In other words, the present (finely tuned) physical conditions were not accidental, as the weak principle assumes, but must have been preordained, and the implications of that - the need for some form of outside influence ('higher intelligence', 'deity', 'God' - the label is unimportant) - to bring that about, must surely be of overriding importance (3.8)?

Yet not one of the 'strong' versions make any reference to this; rather, they have been used to provide meticulous explanations for a wide range of 'cosmological problems'[5], with no indication that these 'solutions' should only be regarded as provisional, pending validation of the strong version of the Carter's principle, and in particular an explanation of just how it's objectives would have been implemented in the first place.

There has to be a fundamentally fallacy in logic or semantics, which allows something which itself cannot be explained, to form the basis of an explanation for something else which cannot otherwise be explained, and though Carter's original proposal made no claims for any sort of causal relationship, it has been modified and extended to such an extent, and to cover such a wide range of different interpretations, that it is not difficult to see why the anthropic principle has become so controversial.

In particular, it was publication of The Anthropic Cosmological Principle, by John Barrow and Frank Tipler[3], which really served to emphasize the distinction between 'weak' and 'strong' versions, in a way which differed significantly from Carter's original concept.

Majority opinion, however, of those who support the principle (and there are many who do not) sides with the weaker version, which in essence states that the 'fine tuned' conditions were inevitable, if the universe was to be suitable for intelligent life able to know and understand it's own background.

10.4 General Considerations:

The anthropic principle reflects the improbability of life, with respect to the physical conditions within the universe. General features and physical properties of the universe are discussed in chapter 2, and those biophysical features which seem specifically suited to the needs of organic life, in chapter 7 (Requirements for Life), though in many cases 'fine tuned' to extremely improbable values in order to be so.

The essence of the anthropic principle is that it highlights this anomaly, and although Carter was not the first to point that out, he was the first to bring it formally to the attention of his professional colleagues. This was in a paper presented appropriately at a major international symposium to celebrate the 500th anniversary of the birth of Copernicus, held in the ancient university city of Krakow, Copernicus's alma mater, and also the first university in Europe to establish a chair of Astronomy.

Basically, Carter disagreed with one interpretation of the Copernican principle, which emphasizes 'uniformity' in general throughout the entire universe, and specifically that human beings do not occupy a privileged position within it.

Originally, this simply reflected the 'insignificance' of earth, within a universe which might even be infinite. It is 'improbability', however, that makes life unique, and in consequence, to use Carter's own words "inevitably privileged to some extent"[6]. (see also Introduction, p2)

Nevertheless, in so far as our knowledge of the universe is incomplete, based as it is on only a 4% sample - the visible observable universe (2.6) - it is impossible to predict weather the Copernican principle would be equally valid if the remaining 96% were taken into account, but certainly few scientist would normally accept such a tiny portion as an adequate representative sample, especially if the remainder could never be seen to confirm it.

What the effects might be if the remainder was taken into account is impossible to say, but with such a large difference, 'butterfly' effects could certainly not be excluded, in which quite small variation in the visible universe could enlarge over time, and eventually become significant elsewhere.

For instance, if the fine structure constant was different by as little as 4% from it's present value, stellar fusion would not have produced carbon, and organic life would never have formed; and even if small amount were possible, from carbon produced in other ways, the universe would have been too cold for it to survive. Either way, without life in the universe, there would have been no one there to know about it[8].

The anthropic principle relates the physical properties of the universe to the presence of life, but although, so far as we know, life exists only on earth, the significance of the principle is 'global', and would take into account the existence of life wherever that might be, whether we knew about it or not. It is important to note however, that 'fine tuning' applies only to carbon based life, and although Barrow and Tipler, no doubt for that reason, restrict their modified 'weak' version (10.2) of Carter's principle to organic life only, there is nothing in the laws of nature, or the universe as it now exists, which would preclude other forms of life, for example based on silicon, and we look at these possibilities in chapter 14.5.

Should any of these exist, however, we would have no way of knowing how successful they might have been in establishing themselves; but one property common to all forms of life is the propensity to 'change', for the difference between 'animate' and 'inanimate' is the difference between 'stasis' and 'progress', and the ability 'spontaneously to change and evolve', is probably as good a working definition of life as any other.

Nevertheless, organic life needs the environment of a planet in which do so, and we now know that there are millions of 'earth like' planets in our own galaxy alone. Privilege implies benefits and advantages, however, and the ability for life to from on other suitable planets, in the same way that it did on earth, is the least one might expect. Having become established, it would then change and progress, but there are too many uncertainties in the 13 billion years of our own past (extinction events, for example - 6.5) to predict how life elsewhere might evolve.

However, perhaps some life forms may never progress to become 'advanced', while others have yet to do so, and that may be among the reasons why we have not yet been able to make contact. Barrow and Tipler, on the other hand, questioning some of the more optimistic estimates for the frequency of life, and using different methods, came to the conclusion that it was actually extremely uncommon - much less than one civilization per galaxy[3].

Nevertheless, with respect to any form of extraterrestrial life, until we have definite confirmation, one way or the other, it is difficult to place human life in any wider context of 'status'. That may seem unimportant, but 'uncertainty' is not something which scientists feel at ease with, especially those working at the forefront of knowledge and technology, and we might certainly learn a great deal if we ever did make contact with a more advanced species.

10.5 Interpretation:

It is very difficult to rationalize all versions of Carter's principle, to be consistent and compatible with one another; ambiguity is too open to misinterpretation; conflicting views are difficult to reconcile; individual versions are different by intent; interpretation by logic can be very different from that based on probability; philosophical and scientific approaches can be semantically unrecognizable, while many simply disagree with the 'anthropic concept' altogether.

The principle as formulated by Carter initially, was in response to the observation, which many would regard as a self evident, that:

> the laws and parameters of the Universe have values consistent
> with conditions for life, rather than values that would not be
> consistent with life.

and

> that this is necessary, because if life was not possible, then no living thing would be there to observe it, and thus no one would known about it[2].

while Roger Penrose, adds yet further clarification:

> That the argument explains why the conditions happen to be just right for the existence of (intelligent) life here at the present time, because if they were not just right, then we would not be here, but somewhere else[7].

It is logic such as that, which permeates all levels of discussion of anthropic philosophy, which make it difficult to understand why so many prominent top cosmologists, among others, genuinely see this as a significant and original contribution to our understanding of life and the universe.

In so far as the anthropic principle comes in two distinct versions, it is not easy to generalize, but nevertheless, it is mainly in it's 'strong' form that explanations have been forthcoming with respect to a variety of cosmological problems. Barrow and Tipler have a monopoly in this respect, and their book, The Anthropic Cosmological Principle, covers a wide range of material. Indeed, john Wheeler in his preface, has likened chapter 6 in particular, to "one of the best short reviews of cosmology ever published".

Other reviews have been less complimentary. Martin Gardner, for example, who clearly disagreed with much of it, described the Final Anthropic Principle, in terms of a supplementary Completely Ridiculous Anthropic Principle (CRAP), and indeed not many books have received quite such a mixed response as this one. Nevertheless, by subtly 'moving the goal posts' they opened up a whole new approach to some of the most intransigent problems in cosmology today, though many who regarded the strong principle as unsustainable, likewise disagreed with these explanations too.

There is a wide variation in standards and emphasis throughout 'anthropic literature' in general, and unfortunately a high reliance on what many would regard as tautologies, truisms, statements of the obvious, or self evident facts confirming each other, and no matter how plausible such arguments may otherwise seem to be, any assertion in which logic has to be unraveled to get to the meaning, is questionable to say the least.

One of the most apt description of the Anthropic Principle nevertheless, calls it 'the central oddity of the universe', and illustrates that with the following example (which also bears out some of the author's earlier comments regarding how statements are 'worded' and 'expressed'):

> The Universe is very old and very large, while man, by comparison, is very new and very young - only a tiny blip on the eternity of space and time. Yet the universe is only very old and large, because we are here to say that it is very old and very large, though of course we all know perfectly well that it is true, and always will be, whether we are here or not[9].

One computer scientist has likened the principle to 'finding yourself in a universe compatible with your own existence, where the conditional probability will always be 'one'.

A common criticism of the strong principle, however, is that it tends to obviate the need to look for other physical explanations, while Roger Penrose sees it more in terms of an excuse, invoked by theorists whenever they do not have any better way to explain their observations[7].

Of the many descriptions and comparisons to be found throughout the literature, one of the most succinct, and possibly appropriate, simply states that the weak principle can be described as a tautology, which neither explains or predicts anything we did not know already, and the strong principle, even more briefly, as little more than unsupported speculation[9].

10.6 Uses:

The weak anthropic principle is a statement of fact; the strong principle, of fact and potential, for irrespective of the ambiguity between the two 'strong' versions, they both imply some form of 'outside influence'. Initially, this was instrumental in determining conditions which would ensure the eventual presence of life; neither version, however, is otherwise in any way specific, and beyond the certainty of our own existence, together with the knowledge that it was not accidental, there is no outward evidence of intervention.

Many of the individual 'coincidence', collectively now embraced under the umbrella of 'fine tuning', have been known about for decades, yet in so far as their relevance was not always clear, or in the absence of obvious explanations in terms of 'known science', it was all too easy to ignore them, as simply further evidence of God's handiwork.

Escalating scientific knowledge, however, over the past half century has changed our attitude entirely, with the need to invoke God (in any form) diminishing in proportion to progress in scientific accomplishments, and with it man's whole philosophy towards the mysteries of nature, from placid acceptance that some things might never be explained, to quite the reverse - that science could explain everything (given time).

The anthropic principle in a sense filled a void that no one quite noticed, by introducing an entirely new philosophical concept of an explanation:

> that things could only happen if it was right for them to happen, because if it was not right, there would be no observers there to know about it.

To most people, just logic and common sense; yet nevertheless has somehow came to be seen in the context of an 'explanation', and thereafter, part of the way in which anthropic philosophy expresses itself. Only the strong principle, however, in which life is seen as imperative, has been used in this way, though some would even see it as the reason why the universe itself exists, while in practice, there are two different levels:

Barrow and Tipler are the archetypal academic approach, and cover a vast range of topics in their massive 700 page tome, The Anthropic Cosmological Prindple[3]. With a skillful blend of mathematics and logic, their arguments are hard to fault, though one is left nonetheless with a feeling that in so far as these ultimately rest on what many would regard as a 'logical fallacy', the SAP, they can be questioned; in short, if one rejects the premise, what follows is difficult to sustain.

More common, is the subjective approach, based on logic alone, tautologies and self evident truisms, usually sequentially related, as an explanatory 'bridge' between an 'outcome' and a 'cause', but in fact largely common sense. Three topical examples should be sufficient to make this clear:

Carbon resonance (7.7) - so improbable that it was both unknown and unsuspected, until Fred Hole deduced that it was inevitable, by the following (anthropic) reasoning:

Elements were known to be synthesized within the core of stars, but in 1954, the production of carbon could not be explained, because the essential precursor, beryllium-8, formed by the union of 2 alpha particles, was unstable, and disintegrated immediately, before it could capture a third alpha particle to form the nucleus of a carbon-12 atom.

Holye had made an extensive study of nucleosynthesis within stars, yet had been unable to account for carbon.

Nevertheless, since life was universal, and life depended on carbon, there must be a way, despite the instability of beryllium-8, and all odds to the contrary, for carbon atoms to form, otherwise life would have been impossible. Hoyle therefore concluded that the only way this could happen would be if the carbon-12 nucleus possessed a very specific resonance level (vibration frequency), which would allow a third alpha particle to be captured (triple alpha process), so that instead of decaying, the beryllium-8 nucleus would convert to a stable carbon-12 nucleus.

No such resonance level was know, and the likelihood that it would exist was considered extremely improbable; nevertheless, if life was impossible without it, then it had to exist.

Hoyle had the greatest difficulty convincing anyone else, however, and only after repeated request did he mange to persuade the astrophysicist Willie Fowler to look for it. Fowlers's success is now history, for which he received a Nobel prize, while Hoyle received no formal recognition.

Hoyle's reasoning may seem no more than logic and common sense, though the fact that it became so imperative, is usually attributed to the strong principle.

Ice and Water - Water is essential to life. It is also the only known non-metallic substance which expand on freezing; hence ice is less dense than water, and therefore it floats on the surface.

Without the insulating effect which this provides, large bodies of water would simply freeze solid, and living aquatic organisms could never have survived. Since life began in the watery environment of the prebiotic soup, it could never have done so, had the properties of water been different. (SAP ± logic)

Multiverse (3.6) - a large (or possibly infinite) collection of universes, predicted both by string theory and by chaotic inflation, all of which have different physical properties. Life can only arise in those universes where these properties are compatible with life, i.e. 'fine tuned'; also, life (observers) can only know about a universe if they exist within it. Hence, our universe had to be 'fine tuned', or we would not exist within it, to know about it.

10.7 Destiny:

We discussed the destiny of the material universe in chapter 2.8. In this section we look at the future of life. The anthropic principles says a great deal about the conditions under which life came into being, and in particular with respect to the improbable physical parameters which allowed it to do so. The odds against our being here, however, to 'observe and understand fine tuning', must be astronomical, while the usual explanations (God or chance), do not elaborate on detail.

Physicists like Stephen Hawking, an acknowledged agnostic, believe that hidden within the laws of nature there has to be an explanation somewhere, and have devoted their careers to trying to find it. Nevertheless, Hawking himself clearly felt the need to hedge his bets, when he stated that it did seem likely that the fundamental constants of nature had been 'adjusted', so that we could exist[10].

The realty, however, is that we do exist, and while some would like to link that to the eventual outcome, in a single comprehensive philosophy of 'existence', others see these aspects as separate and unrelated - with the future more dependent on the present than the past. Some even see 'fine tuning' as an Achilles heel, though after 13 billion years, that would seem unlikely.

The anthropic principle, however, has no predictive implications, and indeed there are very few ways which might shed light on what the future holds. There is one possibility, however - contact with a more advanced form of life - and there are still cosmologists, despite the disappointed results of SETI over the past 50 year, who see this as a realistic possibility, at some stage.

If these do exist, they will undoubtedly have done so through the use of technology far in advance of our own, and although that will not influence their vulnerability to the potential consequences of 'fine tuning', in other respects it will maximize their ability to influence the nature and future of their own long term existence.

If we do eventually make contact, there are three possible outcomes - integration, to a common destiny; incompatibility, war and probably our destruction; or continued separation, certainly in space, and possibly also in time, to independent future.

For the present, however, man sees himself as the pinnacle of anthropocentric evolution, though curiously blind nonetheless to the full implications of his own technology, which has made possible 'tasks' and 'standards' inconceivable a generation ago, and revolutionized the practicalities of 'automation'.

Moore's law may seem deceptively encouraging at present, but it's true significance is very much prophetic, for only so long as man retains control of his machines will he also remain the arbiter of his own destiny.

With technology advancing at it's present rate, doubling in sophistication every few years, it can only be a matter of time before these roles insidiously begin to change, with technology gradually replacing man as the dominant 'life form', artificial intelligence taking over the functions of the human mind, and organic life progressively relegated to an insignificance background role (15.8).

Two aspects (common both to man and to machines) hold the key as to how this is likely to come about in practice - speed and memory, both static for man, but currently doubling every 2 ½ years for machines and technology, and as man delegates more and more, so too the 'anthropic privileges' of 'fine tuning' (on a global scale) will come to matter less and less.

As the dominant life form within the universe changes insidiously from 'organic' to 'artificial', so too will the criteria for long term survival, from those suited to 'carbon production and organic life' to the much less stringent needs of 'inorganic existence', machines and technology, which by then will long since have been fully automated, to remove man altogether from any position of influence.

Such a scenario would be compatible with the weak anthropic principle, and a fine tuned universe arising either by chance, or as part of a multiverse (3.6). The future of life, irrespective of whether it is 'rare' or 'common', should then depend on the stability of the 'tuning', where change in any of life's 'critical requirements' (7.2) could alter it's future irrevocably.

There is another possibility, however, consistent with the strong anthropic principle - that 'fine tuning' was preordained, by some form of intelligent entity, but not belonging to the 4 dimensional of this universe. In these circumstances, the future would no longer depend on random evolution or pure chance, but on whatever future was preordained from the outset.

References (Chapter 10):

1. L. J. Henderson, The Fitness of the Environment (Macmillan, New York, 1913).
2. Wikipedia article (18 August 2013), Anthropic Principle (Variants).
3. J.D. Barrow, F. Tipler, The Anthropic Cosmological Principle (Oxford University Press, 1988).
4. F. Tipler, The Physics of Immortality (DoubleDay, 1994).
5. Wikipedia article (28 August 2013), Anthropic Principle (Cosmological problems).
6. B, Carter, Large Number Coincidences and the Anthropic Principle in Cosmology. (IAU Symposium 63:1974).
7. R. Penrose, The Emperor's New clothes (Oxford University press, 1989).
8. J. D. Barrow, Cosmology, Life, and the Anthropic Principle (Annals of the New York Academy of Sciences, 2001, 950 (1): 139–153).
9. M. Frayn, The Human Touch (Faber & Faber, 1984).
10. S. Hawking, The universe in a Nutshell (London: Bantum Press, 2001).

PART 111

SPECULATION

(Chapters 11 - 15)

Beyond SETI
Aliens Past and Present
Rationale for Contact
Man or Machine?

Chapter 11

FERMI'S PARADOX - Establishing Contact

Enrico Fermi ranks among the outstanding physicists of the last century, and one of that select band of scientists known as the "fathers of the atom bomb". He earned his Noble Prize, however, for work on radioactivity, and also contributed to the evolving disciplines of quantum mechanics and particle physics, which together moved science forward, from the macro world of Newton and Einstein, to where all truly fundamental changes take place. It was a great loss to science that he died at such an early age (53)[1].

Such names as Pauli and Teller were regular companions, and the paradox which bears his name is said to have arise from a conversation about the possibilities of 'faster than light travel' (FTL); this led on to consideration of size and complexity of the universe, and a casual 'aside' by Fermi some time later, reputedly said to have been "where are they?"[2], but surprisingly enough, readily understood by his companions as referring to extraterrestrials.

11.1 Introduction:

The sentiments of this classic paradox, first mooted in 1950, are now generally familiar, but more formally can be summarized in these terms:

> Given the size and age of the universe, it would be surprising if earth was the only inhabited planet; not only would one expect life to be common, with at least as many, rather than than fewer, advanced beings than ourselves, but also vastly more sophisticated and revealing technologies - in which case, why do we not know about them?

In today's terms, this would include anything form visiting aliens, to evidence of enhanced radio emissions from other G2 main sequence or red dwarf stars in the Milky Way - which the Kepler space observatory has now shown to be common; indeed, any advanced life forms in our part of the galaxy, could probably have deduced the existence of life in the solar system, through the increase in radio frequency activity, over the past 50 years.

The implications of the paradox have long been taken for granted in the world of science fiction, where extraterrestrials and aliens are a familiar reality, further entrenched into public acceptance by the appearance of 'flying saucers' in 1947, though more cautiously designated unidentified flying object (UFOs), by a 'concerned', though skeptical military, and attributed largely to anecdotal evidence and media hype (14.9).

Nevertheless, irrespective of terminology, a small number of Unidentified Arial Phenomena (the preferred title today)[3] clearly cannot be so readily dismissed, and pose genuine difficulties to scientific orthodoxy - which by all accounts are taken seriously, albeit not openly.

The reality behind Fermi's original question is even more plausible today, and very few scientists would now dispute the probability of life, in some form, existing elsewhere. What matters even more, however, is the 'level' of that life, and certainly all practical emphasis, so far as the solar system is concerned, assume nothing beyond possible microbial life, such as that currently being looked for by the Curiosity rover on the surface of Mars.

For the universe at large, however, the assumption of 'intelligence' is paramount, and underpins the justification (and funding) for all SETI† projects (11.8,2), since it's inception with project ozma in 1960[4].

Over the years, the paradox has give rise to many scholarly papers and articles, over a wide range of disciplines, from ecology and biology, to astronomy and philosophy, while an in-depth analysis by Michael Hart in 1975, has since become the bedrock for paradox interpretations.

11.2 Background:

It is now over 50 years since Fermi first voiced his concerns, and during that interval, technology has advanced beyond all recognition, and will continue to do so exponentially into the future, together with corresponding implications, not only for ourselves, but even more so for alien technologies which may already be more advanced than our own.

On that basis, one would expect the chances of detecting alien life to increase accordingly, yet in reality nothing has changed - our potentials for success may be greater, but we can draw no conclusions from an unresponsive universe, other than justification to continue looking.

There can be no 'end point, so long as our convictions remain sustained (and fundable), even to the extent, in the words of one latter-day Fermi, "if they won't come to us, we must go to them!", though whether that will ever be achievable must remain very much an open question - until it is.

Hopefully, contact will be established long before that, when it will then only be a matter of time before it becomes 'two-way', and from that point on, technological synergism could change the future beyond all possible predictions at the present time.

In the interim, however, 'uncertainty' will dictate the future - with respect to motivation and feasibility - but only the eventuality of 'proven contact', can guarantee long term prospects, for example with respect to:

- Detection and interaction with extraterrestrial life (ETL).
- Bridging the technology gap between species - whether organic, alternative biochemistry or artificial (life and/or intelligence).

---

† SETI - Search for ExtraTerrestrial Intelligence. A collective name for a number of projects, initiated by the astrophysicist Frank Drake in 1960, to monitor radio frequencies for possible unexplained signals patterns which might be indicative intelligent activity beyond the earth.

- Practicalities of contact - i.e. 'transport', including methods not currently achievable, such as the Alcubierre drive, FTL, time travel and wormholes (13.6).

The overall implications of Fermi's paradox, nevertheless, are even more wide ranging - not only between ourselves and more advanced life forms, but also to embrace the eventuality of a technological singularity (15.3), and the long term future of a universe in which organic life, having fulfilled it's role as the instigator of artificiality, has become redundant, or may even no longer exist.

Hence the conclusion, that the simplicity of Fermi's original question belies the complexities of the answers - for in fact there is no satisfactory answer.

Nevertheless, if we review all aspects of our own life - the improbabilities on which it exists; the relevance of these to the development of life elsewhere; our knowledge of other planetary systems and our advanced SETI technology - we are still left with the inescapable conclusion that we just cannot be the only form of life, within a universe which seems designed specifically to ensure that life can, and must, evolve within it at some stage (10.2).

11.3 Assumptions and Preconceptions:

The arguments behind Fermi's paradox, when it was first propose, rested both on fact, generalities and surmise, while today (and especially since results form the Kepler space observatory over the past 2 years) much of the latter have now been confirmed, with plausible guesstimates replacing 'uncertainty', and a more realistic insight into the difficulties of searching through something which may well be infinite (the universe), for something which may not even exist (life).

Hence, it is solely the balance of probabilities which justify looking for life elsewhere. However, both criteria (on which these are based) and preconceptions, are subjective, and the fact that we exist, and are undoubtedly special and unique (despite the Copernican principle) does not necessarily make it more (or less) likely that similar life forms must exist elsewhere; or that the longer we go on looking, without success, the less likely success becomes; for in the opinion of a panel of scientists at the recent National Space Symposium in Colorado, the chances of finding life elsewhere "will depend a lot on luck".

We can reach a different conclusion, by choosing different criteria, as professors Barrow and Tipler have done, in their authoritative tome The Anthropic Cosmological Principle[5], and when these are applied to intelligent extraterrestrial life, contrary to general opinion, their probability that such life forms exist is extremely low - of the order of one species, or less, per galaxy. However on that basis, since we already exist, that virtually rules out the presence of other intelligent life within the Milky Way, and unfortunately none of our present (or foreseeable) technology would be applicable on an intergalactic basis.

Nevertheless, if different criteria give different results, and none of the criteria used can be shown to be invalid, our whole approach to the frequency of extraterrestrial life, and especially any practical implications relating to it, becomes meaningless. Yet until we have proven evidence that such life does exists, we cannot conclude that it does not, irrespective of the criteria used.

In deciding to look for evidence of life elsewhere, we are already assuming the validity of a number of facts about our own existence, and before we begin to examine the universe for evidence of other life forms, we ought first to reappraise the criteria which justify our own existence, to be sure that these are indeed valid elsewhere.

Life is a dynamic entity - metabolism, growth, evolution - and over time, adaptation, and increasing sophistication of properties such as insight and intelligence, which distinguish human life from other forms.

Implicit in that scenario is the dependence of life on it's surroundings, which in turn must match the requirements of the most highly developed form, irrespective of circumstances - for human life, the 'fine tuned' universe (7.3), no matter how improbable that may be.

Nevertheless, how can we be so sure that within a universe which has no boundaries (3.3) these conditions will apply throughout? and if not, how can we know where or under what conditions other life forms may exist?

Because of expansion, we can compare the values of fundamental constants over the lifetime of the universe, and these confirm the uniformity of the observable universe, which is the only part of it where we can look for life, to a very high degree.

The boundary of that marks the point beyond which expansion of space exceeds the speed of light, and therefore if life did exist in that part of the universe, it would not be able to contact our region of space, either directly or by electromagnetic communication, and likewise (excluding exotic methods of transport - chapter 13) there is no way that we could ever know of their existence.

There are other aspects of the universe which are relevant, however, but which we cannot measure directly, and in particular total mass, which determines whether the geometry of the universe is open or closed, and therefore it's long term future (2.3). We can, however, measure the average density for any given region, which equates with mass, and therefore curvature of spacetime, and prevailing gravitation.

The uniformity of the physical universe is enshrined in the cosmological principle, which states that the universe is the same everywhere (homogeneous) and in every direction (isotropic); this is based not only on direct observation - that physical and chemical reactions are the same wherever they take place - but also the theoretical predictions of general relativity, to which so far no exceptions have ever been found.

A more philosophical view of 'uniformity' comes from the Copernican principle, which initially established the sun as the center of the solar system, replacing the earth from it's historical geocentric role.

By 'demoting' the earth in this way, however, it's importance within the universe was likewise diminished, and it soon came to be accepted that the earth was not special in any way, and certainly not the center of the universe - which the cosmological principle supported.

Nevertheless, on that basis, 'life' must be equally nondescript, and this was later formalized by the statement that;

"humans are not privileged observers of the universe"[6]

That is a significant change of emphasis, however, which does not necessarily follow from the first assertion, and we would disagree with it anyway. So long as man is the only observer of the universe, he will always be privalidged, and his views and 'observations' sacrosanct, for as long as there are no other 'observers' to confirm or dispute them.

Hence, the amended Copernican principle is tantamount to accepting that intelligent extraterrestrial life must exist at some stage in the lifetime of the universe (strong anthropic principle, 10.2) - though not necessarily in our part of the universe†, or even concurrently in time.

---

† Life could even exist beyond the cosmic horizon, which delineates the observable universe, in which case, for the reasons noted above (barring the exotic) there is no way that either of us can ever be aware of the other's existence.

It also highlights an issue of practical relevance. Instinctively, we find it difficult to accept that we might be the only form of intelligent life, and, ironically perhaps, it does not seem as though 50 years of unsuccessful SETI has diminished this intuitive conviction - if anything, quite the reverse, for man is determined to succeed - eventually - in his quest to make contact (in some form or other - chapters 12 and 13) with extraterrestrial life.

As a species, we have now managed to convince ourselves, irrespective of Fermi, that extraterrestrial life doe exist, in such a large and 'receptive' universe, and for that reason, all findings must be subjected to meticulous scrutiny, for man is now instinctively prejudiced, and no longer an objective observer of his own universe..

11.4 Prevalence of Life:

Nevertheless, this remains very much an open question, for as yet, in spite of years of searching, we still do not have one shred of evidence that life exists elsewhere. Fermi's paradox, in fact, simply expresses the same deep rooted instinctive conviction referred to in the previous paragraph, and no one would questions the 'reasons' for that; 'justification', however, is another matter.

We cannot deny the reality of our own existence, yet statistically we are lucky to exist at all; and though we can argue that the reasons we do exist (fine tuning) apply equally to the universe as a whole, that assertion, in turn, rest on first hand knowledge of only 4% of the known universe, while the 'unknown' remainder could well be infinite.

It might also be argued that if, as could very well be the case, life ultimately reflects the influence of some 'outside intelligence' (3.8), then logic and statistics count for nothing; while in an infinite universe, 'prevalence' becomes meaningless, because 'reality' (in our case the 'observable' universe) could never be more than a insignificant part of the whole.

Hence, the question 'how common is life' has a great many qualifications, and aside from curiosity, is far less important than 'quality'. Yet 'intelligence', which that implies, is potentially misleading. Only a tiny fraction of terrestrial life is intelligent, and if that proportion also applies to life forms elsewhere, then only an equally minuscule fraction of that will ever be accessible to SETI, while a correspondingly large proportion of lesser life forms could be saturating the universe, and we might know nothing about it.

Unlike the Fermi dilemma, the reasons for this are obvious, but nevertheless bring reality much nearer to home, and it could be that our first (and possibly only) evidence for life elsewhere, will be found in the solar system.

The hypothetical question of prevalence remains, however, and a variety of theoretical estimates have now been made, which take into account both the limited factual data we possess, and the principle of mediocrity - that the earth is in no way special, but merely a typical planet, subject to the same laws and consequences as any other planet, belonging to any other star system (page 1)

There are estimated to be about 70 sextillion stars in the visible universe, and between 200 and 400 billion of these within our own local Milky Way galaxy[7]. Of the latter, many have planetary systems, while earth-like planets within the habitable zones of their parent star, may also be common.

Nevertheless, such estimates tell us little about life itself. These figures have only become available in the past year, and while they certainly change our views regarding planetary systems and planets themselves, we need to know a great deal more about local factors (chapter 7), before we can extrapolate further regarding suitability for life, and it could be some time before such detailed information becomes available.

As they stand, such figures simply confirm the likely suitability for life to exist, but that does not necessarily equate with whether it does or not, and they are difficult to reconcile with Barrow and Tipler's estimate, for example, of not more than one civilization in the Milky Way galaxy.

They tell us nothing about Fermi's paradox either, except that what started out as an enigma rather than a paradox, is beginning to look more like a dilemma, for it is just as difficult to accept that life is common, yet somehow concealed beyond detection by any means at our disposal, as it is to accept that we are the only living beings in an infinitely large universe.

11.5 Detection of Life:

At the present time, there are two ways in which we can look for evidence of life elsewhere: SETI (11.8,2) which aims to detect intelligent life directly, by looking for evidence of electromagnetic signals which would be a likely feature of any advanced civilization; and by looking for physical evidence of primitive life in the past history of the solar system, and consistent with our understanding of planetary ecology.

Direct contact with planetary surfaces, which has been developing over the past few decades, is still neither reliable or cost effective, and although there are currently two surface rovers deployed successfully on Mars, overall, 45% of all Mars missions have been failures.

A much simpler and more accessible source of essentially similar specimen material is meteorites, which are not only common, but we can now trace their origins and background, with considerable accuracy.

The Allan Hills Martian meteorite (9.4) is the classic example, and illustrates only too well just how difficult it is turning out to be, to authenticate a possible sample of life, to the satisfaction of a very diverse cross section of disciplines, for the physical nature and composition of micro-fossils is such, that distinguishing them unambiguously from artifacts can be difficult if not impossible.

Comparable specimens obtained directly on the surface of Mars, on the other hand, could be even harder to identify. The on- board facilities of the 'robotic rovers' are inevitably limited in the analysis they can carry out, while 'sample and return' missions are still some time in the future.

Nevertheless, within the next few decade, manned missions may eventually become possible, and prospective candidates are already undergoing training. This may facilitate the collection of specimens, but problems of confirmation will always be there.

By contrast, other sources of information, e.g. from comets and interstellar gas clouds (9.4), provide only indirect evidence of biochemistry and organic compounds such as amino acids and RNA, but nothing relevant to confirming the existence of life itself. The fact that the universe shows evidence of underlying 'higher intelligence' is also important, and though commonly acknowledged, little attempt is ever made to incorporate it constructively into relevant arguments such as these (3.8).

In the long term, however, outcomes and consequences may have less to do with 'scientific logic', and more to do with 'intelligent design', while 'fine tuning' which in the past has seemed so important, may actually signify little more than 'suitability', with terrestrial life simply a prototype, and life elsewhere in the universe merely a 'foreseeable eventuality', which may not even come about.

In that respect, 'time' becomes an important parameter, such that civilizations could easily come and go, without ever overlapping, while once interstellar travel becomes possible, civilizations will migrated, and eventually expand to colonize their galaxy within a few tens of millions of years[8].

Existing civilizations by that time would have diversified beyond all recognition, though always leaving a 'loyal core' of ancestors indigenous to their home planet, and with up to 5 billion years, for example, where earth is concerned, before the sun would expand to become red giant, compelling them to move away.

Over that interval of time, Fermi's paradox would long since have resolved, with those beings belonging to the distant future, and now colonizing the universe, themselves the extra- terrestrial life that so obsessed their own ancestors in the remotely distant past. It was then that the question of whether such life existed or not was first raised, and the possibility of detecting it became the incentive for the technology which finally brought it about!

SETI was the first organized attempt to establish the existence of extraterrestrial life, on the assumption that any form of advanced intelligence would likely to be using radio signals for communication in much he same way as we do. It seemed simple and straightforward at the time, using a worldwide network of interlinked radio telescopes, yet after 50 years of continual monitoring, not one confirmed signal has yet been obtained.

There have been many 'false positives' however, all subsequently identified as natural, for example the precise regularity of most pulsar signals, and although data is routinely analyzed by computer, a great deal of human input is necessary too.

Terrestrial TV and radio signals are now such that their collective effect would have increased the solar electromagnetic 'fingerprint', to the extent that other advanced life forms would recognize as signs of 'intelligence'; however, such diffuse signals rapidly diminish with distance, and would not be discernable beyond a few light years.

The situation is very different if signals are focal, rather than spread out, and in theory a suitably powerful laser beam (11.8,2) could easily carry a substantial distance across the universe, and would be readily identifiable provided it could be picked up; however, the more 'focal' a signal is, the narrower the beam, and the harder it is to locate.

Hence, SETI is little more than a hugely sophisticated radio receiver, but it should have been able to pick up extraterrestrial signals long before now, and the fact that it has not done so, is one of the great puzzles in cosmology today

11.6 Physical Evidence of Human Life:

Man began his physical exploration of space on 4th October 1957,when Sputnik 1 was successfully placed in low earth orbit, and space exploration has escalated ever since.

A list of major landmark projects can be found in chapter 5.12, the great majority being artificial satellites and space probes, while between 1957 and the present time, there have been a total of 302 manned space mission,

As an inevitable consequence of this activity, a great deal of physical material is now scattered throughout the solar system, and there are a variety of ways in which this could come to the attention of any suitably advanced forms of extraterrestrial life.

We have assumed that a proportion of these are likely to be more technologically advanced than ourselves, in which case locating such material may well be within their means, depending on it's exact nature, and we can categorized that as follows:

1) Space debris[9]: This includes old satellite fragments, spent rocket stages, and general debris from mutual collision and corrosion.

Although much of this is only centimeters in size, given the velocity of objects in orbit, it still poses a very real collision risk to operational spacecraft, and careful tracking records have to be kept: in 2009 for example, there were 19,000 pieces larger than 5cms, and 300,000 smaller than 1 cm, at altitudes of about 110 miles.

Most operational orbits, however, are well above that, with the ISS for example at 260 miles altitude, and the popular geostationary orbit at 22,000 miles.

2) Artificial satellites: These average a few meters in size, and serve a multitude of purposes. Currently about 3,600 are still in orbit, at altitudes between 80 miles (low earth orbit) to 22,000 miles (geostationary).

3) Scientific structures and equipment: The great majority of these are dedicated satellites in geostationary orbit, e.g. communication, TV, navigation, military etc, while other large discreet objects include the Hubble Space telescope - 43 ft long, it orbits the earth once every 96 minutes, at an altitude of 347 miles; and the International Space Station (ISS) which is the largest object in the local vicinity of earth. It measuring 240 x 355 x 66 feet, and completes one orbit every 92 minutes, at an altitude of 260 miles.

4) Artificial objects on planet surfaces: Only those on the exposed surfaces of the moon and Mars, where there is no atmospheric obscuration, could theoretically be viewed from a distance. These include, on the Moon, 5 third stage Saturn V rockets, laser ranging equipment, and overall, 187 tons of material; and on Mars, residual artifacts include remnants of 3 crashed space probes, and the Opportunity (MER-B) and Curiosity rovers, both still in operation as of June 2014.

At present, even the most powerful instruments on earth cannot detect such artifacts, and it would be highly unlikely that alien technology would ever be able to do so either - from a distance. However, flybys or orbiting probes can locate surface objects quite easily, and a suitably advanced civilizations the might be able to do the same, with a Bracewell probe (11.8,4).

There are two other categories of artifact, however, which are not arbitrary, and some of which include 'designer information' specifically intended to make our presence know to any life forms which might eventually come across them:

5) Artificial objects escaping from the solar system[10]: A number of deep space planetary probes, having completed their intended missions, have passed beyond the heliopause, into interstellar space.

Pioneer 10 - Launched in 1972, and now heading in the direction of Aldebaran in Taurus.

Pioneer 11 - Launched in 1973 is now heading toward the constellation of Aquila, and barring accidents, will pass near one of the stars in that constellation in about 4 million years.

Voyager 1 - Launched in September 1977, passed the heliopause on 25 August 2011 to enter interstellar space.

Voyager 2 - Launched in August 1977, passed Neptune in 1989, and is now heading outward, though has not yet passed through the heliopause.

New Horizons - Launched in 2006, will make a flyby of Pluto in 2015; it's third stage, a STAR-48 booster, is on a similar escape trajectory from the Solar System, and will probably cross Pluto's orbit in late 2015.

6) Designer artifacts confirming intelligent life: There are two of these in physical form, and one as a radio signal:

(a) The Voyager 1 space probe, which has now exited the solar system into interstellar space, is not heading toward any particular star, but will pass within 1.6 light-years of the star Gliese 445, in the constellation Camelopardalis, in about 40,000 years.

Optimistically some might say, it carries a gold-plated audio-visual disc, which would identify our presence to any intelligent alien life who may come across it in the future.

The disc includes a photograph of earth, and of human beings, a spoken greeting from the American President (Jimmy Carter), a range of scientific information, recorded sounds, including a whale, a baby crying, waves breaking on the shore, and of selected classical and modern music, as well as a brief greetings message in 55 different languages.

(b) The Pioneer Plaque. Both Pioneer 10 and 11 carried a gold-anodized aluminum plaque, measuring 6" x 9", and attached to the antenna support struts to provide shielding from interstellar dust. The plaque contained information which would clearly identify human life, included nude male and female figures, and a variety of symbolic information designed to identify the origin of the spacecraft - for example the position of the sun with respect to certain pulsars, whose unambiguous signal frequencies were known with great accuracy.

(c) The Arecibo message was a frequency modulated radio signal, transmitted from the Arecibo radio telescope in Puerto Rico on 16 November 1974. It was aimed at the globular star cluster M13 some

25,000 light years away, which was the largest and closest suitable 'target' at the time.

The message consisted of 1,679 binary digits, approximately 210 bytes, transmitted at a frequency of 2,380 MHZ, and encoded 7 groups of basic biophysical information:

- The numbers one 1 - 10.
- The atomic numbers of the elements which comprise DNA.
- The formula for the sugars and bases in the nucleotides of DNA
- A graphic of the double helix structure of DNA.
- A graphic figure of a human, and the human population of Earth.
- A graphic of the Solar System, identifying the position of earth.
- A graphic description of the Arecibo radio telescope.

In reality, the message was no more than a demonstration of man's level of technology, and could never have served any practical purpose. The theoretical round-trip signal time would have taken 50,000 years, though due to the 'proper motion' of stars comprising the M13 cluster, it would no longer be in it's original position anyway, by the time the signal arrived.

Nevertheless, there must always be the possibility that some artifact or information, disseminated from this planet into the universe at large, might at some time in the future be intercepted and understood; and with all the uncertainties of technology, of time and of space, we can never preclude the possibility of a response.

11.7 The Drake equation:

In 1959 an article entitled 'Searching for Interstellar Communications', was published in Nature[11] which suggested that the large radio dishes now in use, might be sensitive enough to pick up signals coming from alien civilizations, who would likely be using the same methods of communication as we do.

This was the first practical suggestion by which extraterrestrial life might be detected, and in 1961 the astrophysicist Frank Drake initiated a systematic search for evidence of such activity, using the 25 meter dish at the National Radio Astronomy Observatory at Green Bank, Virginia (project Ozma[12]). Two nearby sun-like stars, Epsilon Eridani and Tau

Ceti, were scanned for 6 hours daily, using frequencies close to the common 21cm wavelength of neutral hydrogen, over a 4 month period.

The experiment was unsuccessful, but to get some idea of what sort of signals might have been expected, Drake went on to made a list of all the parameters that would have to be taken into account in order to detect evidence of intelligent life associated with some other star system; if all of these factors were then multiplied together, that would be the number of 'communicating' civilization (N) in our galaxy.

The equation had the following form[13] [14]:

$$N = R\% \cdot f_p \cdot n_e \cdot f_l \cdot f_i \cdot f_c \cdot L$$

where:

R% = average rate of star formation fp = fraction of stars with planets
n_e = average number of planets per star able to support life
f_l = fraction of planets able to support life that actually do.
f_i = fraction of planets with life that develop intelligent life.
f_c = fraction of civilizations with detectable technology.
L = Length of time such technology could have been detected

There were many drawbacks, however, and the equation is essentially academic, and of little practical value with respect to providing a realistic answer.

With the exception of R%, for example, none of the terms are known with any accuracy, and most cannot even be estimated in any true sense. The equation makes no allowance for relative importance, placing equal 'weight' on all factors, and it does not specify any particular life form. Factors n_e and f_l, for example, must apply to organic life, while other factors would apply to 'intelligence', which could include different forms of life.

Hence, aside from numerical difficulties, ambiguities such as these make the equation essentially meaningless, and until we have proven evidence that life elsewhere does exist, we cannot make assumptions as to it's nature, and especially not based on 'instinct', which has hardly proved helpful so far as 'presence' is concerned.

All options must remain open until we have evidence to the contrary, and the most useful purpose of Drake's equation has been to clarify the many factors which do have to be taken into account, in considering the question of intelligent life elsewhere.

The equation has since been modified, however, for example to take into account 'contact' between alien intelligences, expansion of civilizations to colonize other stars stems, civilization extinction, and colonization of a give planet more than once, by successive civilizations over suitably long intervals of time.

The format of the equation has not changed, but all parameters have now been revised, especially in the light of recent Kepler findings, and considered individually these are more likely to be helpful, than trying to take all of them into account at the same time.

Initially, Drake made the following estimates for each term:

R%= 1 per year (average rate of star formation).
f_p = 1/5 - ½ of all stars (fraction of stars with planets).
n_e = 1 - 5 (planets per star able to support life).
f_l = 1 (100% of planets will support life).
f_i = 1 (100% of these will develop intelligent life).
f_c = 10 - 20% (civilizations able to communicate).
L = 103 - 108 (Length of time communications detectable.)

It is not clear why these particular numbers were chosen, but in the words of one skeptic "the values attributed to each factor in this equation tell more about a person's beliefs than about scientific facts".

Nevertheless, Drake made a number of provisional calculations, and came up with a value of between 1000 and 100,000,000, for the number of communicating civilizations in this galaxy.

Reasons for choosing different values, range from those which assume the principle of mediocrity (Introduction, p1) at one extreme, to those based on the rare earth hypotheses - an argument introduced in 2001, which noted that intelligent life on earth required an almost unique combination of astrophysical and geological background circumstances, and therefore any corresponding level of extraterrestrial life would require a similar earth-like planet, with equally rare conditions, and that such planets would probably be extremely uncommon, if they existed at all.

Hence, if we solve the Drake equation

$$N = R\% \times f_p \times (n_e \times f_l) \times f_i \times f_c \times L$$

for each of theses two extremes we get:

- Lowest estimates (rare earth hypothesis, where n_e x f_1 = 10^{-11}, or one inhabited planet only in the galaxy), which gives:

$$N = 7 \times 0.4 \times (10^{-11}) \times 10^{-9} \times 0.1 \times 304 = 8 \times 10^{-20}$$

- Highest estimates (principle of mediocrity), giving:

$$N = 7 \times 1 \times 0.2 \times 0.13 \times 1 \times 0.2 \times 10^9 = 34.6 \text{ million}$$

Clearly, figures of this sort are quite meaningless, and the science fiction author, Michael Crichton, in a lecture at Caltech in 2003, made this very clear when he said:

> "the problem is that none of the terms can be known, and most cannot even be estimated. The only way to work the equation is to fill in with guesses, and on that basis the Drake equation can have any value from "billions and billions" to zero.. ....and speaking precisely, the Drake equation is literally meaningless".

By contrast, however, the astronomer Seth Shostak has stated that he expects to get conclusive evidence of alien contact between 2020 and 2025[15], based on the Drake equation[13], and that might be good news for SETI@home (11.8,2), despite its fourteen unsuccessful years so far.

It is also easy to forget that Drake originally formulated his equation only as an agenda for discussion at the Green Bank conference[16], which he was organizing at that time - speculation about unknown parameters intended only to stimulate interest in how best to proceed in the future, rather than a formula for answers.

11.8 Facts and evidence:

The facts and information necessary to understand Fermi's paradox, and comprising much of the discussion in this book, can be grouped under four categories:

(1) Astronomy: This is the historical basis for everything we know about the universe, and only within the past 100 years has it diversified into the many specialist disciplines which now exist.

Strange lights, objects in the sky, and many related phenomena have been recorded throughout human history, many now clearly common celestial phenomena, some less common, and that small illusive residue that still cannot be fully explained, and will always be attributed to extraterrestrials by somebody, until they are.

At a more professional level, natural phenomena which raised genuine initial doubts, included pulsars, because of the precise repetition of their signals, and Seyfert galaxies, where their enormous energy output suggested some form of industrial accidents, before the true explanation became clear.

Other possibilities for advanced civilizations might include large scale use of solar power, altering a planets light curve, or asteroid mining, which could change the appearance of 'debris discs' around stars.

(2) SETI: Distant observers of the Solar System would be able to detect unusually intense radio activity for a G2 main sequence star, due to terrestrial TV and radio transmissions. Hence, theoretically we could do the same, if similar alien signals existed, though only within a few light years of the sun, where very few suitable planets have yet been identified.

Radio signals penetrate the earth's atmosphere quite readily, and artificial signals, which characteristically would be repetitive, with a narrow band width, would not be difficult to identify.

Many SETI project now exist, since project Ozma closed down, including Cyclops, Phoenix, SERENDIP and the Allen Telescope Array, all basically monitoring a range of frequencies continuously 24 hours a day. Only one suspicious signal has ever been found, however, over 50 years of scanning - the Wow! signal (9.8), picked up by The Big Ear radio telescope; however, this instrument only looks at each point for 72 seconds, and re-examinations of the same spot later found nothing.

Large numbers of radio telescopes are also now connected together and synchronized, world wide, allowing not only interferometry, but also greatly increasing range and efficiency.

As originally conceived, the objectives of SETI were to search for evidence of life up to a distance of 1000 light years. Nevertheless, although instruments and technology have advanced progressively, there are many technical assumptions underlying SETI analysis, which is not always straight forward, and we can never be sure that something significant might not have been missed, or perhaps more likely misinterpreted.

SETI@home is an internet project set up in 1999, to increase computing power, by linking up a network of home computers, using customized software to analyze selected data fed in from the Arecibo radio telescope. With 145,000 active users, in 233 different countries, the project has now logged over 2 million years of aggregated computing time since it's inception, easily matching the power of even the largest supercomputer[17].

By 2010, SETI@home had scanned every point in its chosen frequency at least 3 times, over 67% of the observable sky.

Optical SETI is a theoretical alternative to radio waves, based on the assumption that alien civilizations might be using powerful lasers for long range or even interstellar communication. A preliminary study using a high-energy laser and ten-meter diameter mirror, has shown that such a signal, focused into a narrow beam, would appear thousands of times brighter than the Sun, to any distant civilization in the exact line of sight[18].

Provisional optical SETI programs have been in place now since 1998, with 2,500 stars examined in the first year, and the Harvard-Smithsonian group is now building a dedicated all-sky optical survey system using a 72-inch telescope, being set up at the Oak Ridge Observatory in Harvard, Massachusetts.

So far, however, the only laser signal actually received from space, with information encoded on it, came from the International Space Station on 5th June 2011. This was a video message, transmitted at 50 megabits/second, and took only 2½ minutes, compared with over 10 minutes using traditional download methods.

The main difficulties are that unlike radio waves, which spread in all directions, laser beams are highly focused, and therefore extremely difficult to locate, and would also be easily blocked by interstellar dust.

High energy Gamma-ray bursts are common throughout the observable universe, on average about one a day, and so far remain unexplained. Hence, it has been suggested that these could be generated artificially by civilizations which have reached a technological singularity

(15.3). However, such signals do not penetrate the atmosphere, and are not routinely monitored by SETI.

Collectively, such fact and figures are remarkable statistics, as to the huge amount of work that has been carried out to date; but not just impressive, they also highlight ever more clearly, the true extent of the anomaly implicit in Fermi's paradox. Surely we should have detected something by now?

(3) Planetary observation:

Kepler has now identified around 1800 exoplanets beyond the solar system - 1822 planets in 1137 planetary systems, including 467 multiple systems, as of 12 September 2014 and it is estimated that there could be as many as 40 billion earth-sized planets orbiting in the habitable zones of Sun-like stars, and red dwarfs within the Milky Way[19].

Such planets are rarely observed directly, and their existence usually inferred by other means (12.10), when together with information about their parent star, general conclusions as to composition and the planetary environment can also sometimes be made.

Direct evidence of life will be the next stage forward, when it may be possible to detect signature gases (e.g. methane and oxygen, 12.5), or possibly even industrial pollution of a technologically advanced civilization.

(4) Alien Evidence (14.9):

Alien probes, such as a self-replicating Von Neumann probe (15.5) could theoretically explore a galaxy the size of the Milky Way in as little as half a million years. Probes would spread throughout the entire galaxy, and evidence of such surveillance might be found in the Solar System, e.g. in the asteroid belt, which could be used as a potential source of raw material.

Bracewell probes (15.5), specifically sent out by alien civilizations to collect evidence, and return information, but without making actual contact, are another hypothetical suggestion, popular in particular with flying saucer enthusiasts.

Remains of alien artifacts (14.9), either on this planet or elsewhere in the solar system, have been suggested for years, with the very real possibility that, whether by design or otherwise, if they did exist, they might go unrecognized.

In addition, alien technology has long been proposed to account for many of the ancient structures and remains worldwide, for example the Pyramids of Egypt and Stonehenge.

Erich von Däniken, and his best-selling books Chariots of Gods (14.9) and God from Outer Space, popularized belief in alien visitations, as an explanation of many ancient monuments and artifacts - most long since disproved, though he was an honest believer in his own ideas, rather than a crank, and subsequently went on to become a co-founder of the Archaeology, Astronautics and SETI Research Association.

11.9 Difficulties of Resolution:

It is hard to conceive of a more diverse collection of facts, evidence and opinions, than those associated with Fermi's paradox, which started as a simple anomaly - to reconcile strong human intuition with total lack of evidence to back it up.

Numerous scholarly books, papers and articles have been forthcoming over the years, while Stephen Webb's definitive classic, Where is Everybody[20], published in 2002, lists 50 different explanations, covering every possibility aspect in the greatest detail.

For those interested, a portion of this book can be accessed directly, by clicking twice on reference 53 in the Wikipedia Drake Equation article (issue, 5 December 2013). Unfortunately, only the first 39 pages are included, but these make interesting reading nonetheless.

All of this simply reflects the reality, that the whole question has escalated out of all proportion over the 54 years since Fermi uttered just 3 words†, and set facts and supposition at loggerheads ever since, such that 'separating wood from trees' is now very much part of the problem.

There are many more complex issues involved, however, than three simple words might suggest, and we need to look at these in more detail first, before returning to discuss the whole question of extraterrestrial life (chapter 15) and the background factors to that, which might help to explain the anomalies of the paradox itself.

---

†   "Where are they"?[2]

References (Chapter 11):

1. Enrico Fermi Dead, 53 (New York Times. 29 November1954).
2. Google: The fermi Paradox (multiple sites).
3. Project Blue Book - Unidentified flying Objects.
4. Science: Project Ozma (Time, Apr. 18, 1960).
5. J.D. Barrow, J. F. Tipler, The Anthropic Cosmlogical Principle (Oxford University Press, 1988).
6. J. A. Peacock, Cosmological Physics (Cambridge University Press, 1998), p66.
7. A .Craig, Astronomers count the stars (BBC News, July 22, 2003)
8. I. A. Crawford, Where are they? Maybe we are alone in the galaxy after all (Scientific American, July 2000: 38-43.
9. Wikipedia article (16 March 2014), Space Debris.
10. Wikipedia article (30 August 2014), List of artificial Objects Escaping from the Solar System.
11. A. M MacRoberts, The Chances of finding Aliens (Sky & Telescope, 3 June 2009).
12. Google: Early SETI, Project Ozma, the Arecibo Message.
13. SETI Institute: the Drake Equation.
14. Wikipedia article (5 December 2014), Drake equation.
15. S. Shostak, First contact Within 20 Years (Space daily, 22 July 2004)
16. The Drake Equation Revisited: Part I (Astrobiology Magazine. 29 September 2003)
17. SETI@home, Website
18. R. Exers, D. Cullers, J. Billingham, L. Scheffer, SETI 2020, A Roadmap for Search for Extraterrestrial Intelligence (SETI Press 2003, ISBN 0-9666335-3-9).
19. Wikipedia article (25 September 2014), Exoplanets
20. S. Webb, If the Universe Is Teeming with Aliens... Where Is Everybody? (Copernicus Books, 2002, ISBN 0-387-95501-1).

Chapter 12

SEARCHING FOR LIFE

The possibilities of other life forms existing elsewhere in the universe are scattered throughout the myths and legends of history, but there has only been one serious claim to actually have identified alien life, when the astronomer Percival Lowell presented his theory of 'canals' and intelligent life on Mars, in 1895. He was a professional of some standing, nonetheless, and his published findings in 3 beautiful illustrated books[1], did much to popularized the long-held belief that changes in the markings on the Martian surface were evidence of intelligent life[2].

Fifteen years on, however, and a much more powerful telescope at Mount Wilson Observatory allowed closer observation of the structures depicted by Lowell, and revealed irregular geological features, probably resulting from natural erosion.

Nevertheless, there was nothing illogical about Lowell's work, and his findings were based on exactly the same assumption that we are making today - that life elsewhere must be common - though knowing nothing about the real universe, he only had the solar system to go on, and there were certainly visible changes on Mars which had to be explained. It is not usually appreciated, but it was actually Lowell's study of these markings which gave rise to the ideas behind SETI in the first place (12.4)!

Having now explored the solar system in detail, it is easy to be dismissive of Lowell, but in fact the assumptions are still the same today, and all we are doing is applying them to the universe at large. Only time will tell whether they are more justified for us than for him, but the first 50 years are hardly encouraging, and it may be that there are lessons to be learnt from the past.

12.1 Background:

Over that same period of time, and particularly the last 10 years, technology has escalated so rapidly, that the search for planets beyond the solar system, including the identification of potentially habitable planets, has made exo-planetology and astrobiology the fastest growing disciplines in cosmology today.

Nevertheless, it could well have been Fermi, back in the 50s, and his preeminent reputation among colleagues, who first caused professionals to take seriously the whole question of extraterrestrial life, and in particular the possibility of actually being able to find it.

It was 10 years, however, before the astrophysicist Frank Drake took up the challenge, with project ozma (11.7), using a large radio telescope to try to pick up non-random signal patterns, suggestive of intelligence, and possible evidence of extraterrestrial communicating civilizations.

This was at a time when it was assumed that alien life as advanced as ourselves was probably common, and would be using radio communication in just the same way that we do. There were still many skeptics, however, and Drake's greatest achievement may well have been to get funding for such a project in the first place, and the important precedent that went with it.

Electromagnetic radiation is probably the most ubiquitous property of nature, and as an artifact, might be taken to signify 'coming of age', for any form of intelligent life, when it's existence would then be 'in the public domain' and theoretically could no longer be concealed from others.

The reasons why any civilization might want to remain hidden, however, are enigmatic, and simply take us back to Fermi, and all the unanswered questions his paradox brought to light.

Nevertheless, searching for extraterrestrial life is certainly more rational today than it was for Lowell, with a diversity of skills and technology beyond anything that he could ever have imagined. Ultimately, however, human interpretation remains the deciding factor, irrespective of the quality of evidence, and with it the fallibility of subjectivity.

12.2 Options for Search:

Life is manifest in a great variety of way, but from the point of view of simply establishing 'existence', there are only two criteria:

- Identifying a living entity, which at the most basic level, could well be more difficult than it might seems - the status of desert varnish for example (14.9), as to whether it is living or not, has been debated for over 150 years, and is still undecided . At the opposite end of the spectrum, however, intelligent radio signals would be unambiguous.
- Manifestations of life, whether past or existing in the present.

These two possibilities, in turn, decide the methods best suited to searching for life, and the circumstance under which it would most likely be found, which broadly come under 5 categories:

1) Radio, e.g. SETI - listening for extraterrestrial radio signals, indicative of Intelligence and advanced technology.
2) Direct exploration of planet surfaces - ideally for evidence of live microbes or bacteria, but more likely, microfossils. At present carried out by 'robotic rovers' (5.11), though manned exploration of the moon and Mars are likely within a few decades.
3) Meteorites, for similar evidence of possible microfossils (e.g. the Allan Hills meteorite, 9.4), which would lend support to panspermia, or possibly even live organisms, in some form of hibernation or suspended animation. Nevertheless, even confirming possible microfossils is proving extremely difficult to do (11.5).

4) Biochemical evidence, obtained spectroscopically from such diverse sources as planetary atmospheres, exoplanets, comets and interstellar dust and gas clouds. This can never be more than 'suggestive', but is valuable, for example, in assessing potentially suitable environments.

5) Searching for extrasolar planets (exoplanets), and in particular those within habitable zones of their parent star, though again, these are not confirmation, either of life, or even suitability for life. We need to know a great deal more about atmospheric environments first (7.9), and how these might change and evolve over time, in relation to potentials for the origin and evolution of life, before we can draw conclusions. The prebiotic environment on earth, for example, was very different from that in which life subsequently evolved (8.3).

12.3 Rationale:

It is natural for man to see 'human life' as paramount, but a mistake to interpret that beyond his home planet; yet the Copernican principle (5.3) which man himself formulated, and states that life has 'no favored position or status' within the universe, would clearly be meaningless, unless life did exist elsewhere.

Nevertheless, after 50 years of trying, man has so far failed to detect evidence of alien radio signal, which would have confirmed his belief that intelligent extraterrestrial life is common, and maybe his anthropomorphic modesty was premature.

Yet despite these beliefs, man himself has done little to reciprocate, with only one formal radio attempt to make his own presence known†
- the Arecebo message (11.6,6c), with basic information about human life, directed at the globular cluster M13 in Hercules (11.6,6c). With 50,000 years for the round trip, however, response was never the intended purpose, but rather to place evidence of humanity on the blueprint of eternity.

Size, however, is another parameter which man has misinterpreted. The universe may well be infinite, both in space and in time, for the

---

† Other non-radio messages describing human life, have included the Pioneer plaque carried on Pioneer 10 (1972) and Pioneer 11 (1973), and the audio-visual disc carried on Voyager 1(1977) (11.6,5,6).

reality of 'spacetime' precludes one without the other. But that too has little relevance to the search for life.

Any finite property, such as prevalence of life, no matter how large, will always be rare within an infinite environment, and the common terminology of everyday usage cannot apply, to the presence or concepts of life, in a universe such as it is.

Nevertheless, rarity is not necessarily the reason behind Fermi's paradox, which might be rationalized in a number of ways, but is difficult to reconcile with factual evidence regarding 'suitability'- designer universe, fine tuning, geometry, carbon dominance, etc, and therefore a very good reason to go on looking; however, it does not change the reality that we may indeed be the only form of life that exists.

The whole question of existence, however, must ultimately depend only on established facts, and once we accept that reality can only be what we find, rather than what we expect, at least some of the anomalies become easier to understand.

It is a fallacy of logic, nonetheless, to argue that because life exists on this planet, it must necessarily exist elsewhere, and nothing more than wishful thinking. No matter how natural or intuitive that may seem - "surely we cannot be the only one" - it can never be more than that, and there is neither factual or statistical justification to support such an assumption.

None of this affects how we actually go about looking for life, so long as we disregard preconceptions, and in particular the basic supposition we started out with - that life in the universe must be 'common' - because one cannot generalize from a single sample, and 'common' is a meaningless term in any context such as this.

So long as we go on believing it, however, Fermi's paradox will continue to baffle cosmologists; yet there is an equally plausible alternative, compatible both with 'lack of success' and with 'life being common', but which constantly seems to be overlooked - that we have simply not been able to recognize their signals.

Nevertheless, if we look at a selection of our own languages, there are always certain common features with respect to syntax, ordering and repetition, which will still be manifest irrespective of symbolism. All of these were important primary features in the initial deciphering of Egyptian hieroglyphics, for example, while it was largely 'repetition' and 'circular logic' which enabled Bentley Priory to break the WW2 Enigma code.

Hence, if we can program-in all of these principles for the computer analysis of 'noise patterns', then any which do contain 'intelligence' will be subject to the rules of common logic, and therefore with the right algorithm could be rendered meaningful.

Nevertheless, assumption of similarity, at any level, for something so different as an alien intelligence, will always be a gamble,. One would expect 'principles and logic' to be independent of intelligence - provided that intelligence itself cannot change - or could it, depending on biochemistry, for example?

It is all too easy to be dismissive of anything at odds with our own standards and conventions, and 'similarities' will no doubt be a prominent topic for debate, if and when we ever do make contact with an alien species.

12.4 Radio Signals:

Given that we are now committed to the search for life, and ideally 'to make contact', even lack of success so far is no reason to give up, and it still seems likely that radio (as the only means of bridging vast distances across the universe) must remain one of the most likely ways of achieving this objective.

Nevertheless, that need not necessarily be at the present time, and we look at reasons later which could easily explain why: that such signals are not yet there to detect is the obvious one, with absence of aliens themselves the reason for that; but there is another very simple reason, other than unfamiliarity - that our technology (or methodology) is inadequate, or inappropriate.

The latter, for example, might be if the aliens are not, after all, dependent on radio communication, but on some other method - and depending on how advanced they may be, that could be anything from optical or laser-based communication, comparable to our use of fiber optics, to telepathy or ESP.

For these possibilities, however, aliens would have to be either very different from us, or very advanced, and we shall argue later why in fact we do not believe that aliens are likely to be significantly more advanced than we are, and therefore neither of these reasons seem likely for day-to-day routine purposes. In addition, it would be impossible to develop such advanced technology, without an intermediate dependence on radio communication.

As to being 'inadequate', it is difficult to be comprehensive with such a large and diversified project, including so many variables. These range from scanning (where, how long, how often), repeating (routine or 'suspicious' signals only), and programming, to data collection and analysis, with over 100 radio telescopes world wide, and the 145,000 PCs of SETI@home,

Organization on this scale is a mammoth task, while the scientific value of so many instruments and analytical facilities, is to some extent offset by the difficulties of coordinated integration.

There will always be a disparity between the virtually infinite cosmos and the finality of man's potentials, with only 'persistence' to bridge the gap; though any deficiencies in matching 'project' with 'objectives', more likely reflect the nature of the project, than inadequacy of facilities or ability.

We can summarize the essence of all of these quite simply: in searching for alien life, we have to depend solely on our own technology and experience, for simple practical reasons. However, these in turn must assume at least some similarity between us and them, for the project to be practical at all, while the reality is, that by virtue of being 'alien', they will always be different from us; and just how different, in relation to the assumptions we make, is what determines how successful or otherwise, we are likely to be.

Radio astronomy, of which SETI is now perhaps the most active branch, stems from the work of Karyl Jansky, who built the first radio telescope in 1931 - a 'skeletal' array of antennae, mounted on a long horizontal rotating arm, and the first instrument designed to detect and study the longer wave lengths of the electromagnetic spectrum, at the opposite end from those of visible light.

It took many months to identify and classify different natural sources, but when all of these were excluded, Jansky was left with a faint cyclical 'hiss', which was eventually identified as coming from the Milky Way galaxy, and maximum towards it's center in the constellation Sagittarius.

Modern radio astronomy, however, stems from the radar technology of WW2, and studies cosmic radio emissions from natural processes, both thermal and synchrotron, emitting in the long wavelength region of the electromagnetic spectrum. This ranges from millimeters up to about 1 kilometer, and includes signals coming from radio stars and galaxies, supernovae, pulsars, quasars and emission nebulae.

These are matched by an equally wide range of instruments, though basically belonging to one or other of two types: steerable antennae, consisting of a large parabolic dish, commonly between 50-100 meters in diameter, built onto a computer controlled altazimuth mounting; and stationary antennae, embedded in the ground, with a moveable receiver to provide some range of flexibility.

The Arecibo radio telescope, with it's 1,000 ft dish is the largest of these in the West, and also doubles as the largest planetary radar telescope, while the largest individual radio telescope is in Russia, with a circle of radio reflectors, 1,700 ft across, each focused onto a central receiver.

There are now over 100 radio telescopes world wide (including both single dish and interferometric arrays), though only some are active participants in SETI at any one time, and many are dedicated instruments. Examples of these include the Big Ear telescope, using the ubiquitous 21 centimeter line of neutral hydrogen, at the 1420.40 MHz frequency at which the Wow! signal (11.8,2) was detected, the Waterhole, using frequencies from 1,420 to 1,666 MHz, and the Arecibo Observatory, which has a number of receivers, covering the whole 1–10 GHz range.

The earliest suggestion that radio signals might be used to detect or even contact extraterrestrial life, and therefore indirectly gave rise to SETI many years later, arose in relation to Lowell's theory of Martians, in 1895. This had led Nikolas Tesla, who pioneer 'wireless' several years before Marconi, to suggest the following year that his new system of 'electrical transmission' might be used to contact Mars, and even claimed to have detected extraterrestrial signals[3].

A few years later, Marconi himself claimed to have detected signals from Mars[4], while in the 1920s, the US army actually assigned a cryptographer to translate any messages receive from Mars!

In 1955, an article in Scientific America suggested the possibility of scanning the cosmos for radio signals, and Drake's project Ozma may well have been influenced by that, while quite independently the Soviets too were actively trying to detect 'signals from space'.

In 1971, NASA sponsored a study (Project Cyclop) of an array of 1,500 radio dishes, which was never built, but the idea formed the basis for much of the work on SETI which subsequently followed.

Practical developments really began a decade or so later, with project Sentinel in 1983, having 131,000 narrow band channels, followed by project META in 1985, with 8.4 million channels, and by 1995, project BETA was able to process 250 million channels simultaneously. Unfortunately, the 26-meter radio array on which all of these project were based, was later destroyed in a storm, and all three projects have now been superceded.

The Microwave Observatory Program (MOP) was set up in 1992, with government funding, as part of a long term general survey of the sky, together with targeted searches of 800 nearby stars. Unfortunately, this was subsequently ridiculed by U.S. Congress, and cancelled a year later. The project continued without government funding, until taken over by the SETI Institute and privately funded as project Phoenix, to survey 1000 near-by sun- like stars.

Much of SETI is now conducted jointly by professionals and amateurs, as the SETI League[5], though it has been estimated that to keep SETI activities going world wide, would cost around £14 million ($20 million) per year.

Paper SETI, under the auspices of the SETI Institute, is an adjuvant to the radio network, whose task is to design suitable messages, conveying basic scientific or mathematical principles, together with human attributes, that could be sent out to any extraterrestrial, once contact had been successfully established.

The idea stemmed from a workshop, held in Paris in 2003, to bring together a multi-disciplinary group of anthropologists, philosophers, theologians, astronomers, physicists, musicians and artists, to consider the problems and difficulties of communicating with extraterrestrials in a language and syntax that might be intelligible to them. We would also add, an Egyptologist, and the army cryptographer referred to earlier, or his successor, to that list!

12.5 Spectroscopy:

Radio technology, such as SETI, is a 'search tool', and as such has not yet contributed in any true sense, to the search for extraterrestrial life. By contrast, spectroscopy, which studies the interaction between matter and radiation, expressed in terms of wavelength, is perhaps the single discipline of physics which has advanced our knowlege of the universe more than any other, and is already playing an active role with respect to the identity and physical properties of exoplanets (12.10).

When light from an incandescent solid is passed through a prism, it is dispersed into it's constituent colors, and displayed according to wavelength as a continuous strip of color. (continuous spectrum).

When a single element is heated, it emits light of a particular wavelength (color), characteristic of that element. If white light is passed through an incandescent gas containing a specific element (e.g. in the outer layers of a star's atmosphere) that element then absorbs light corresponding to those wavelengths which it would normally emit, and appears instead as a single dark line; if several elements are present, each will absorb light of it's own wavelength, to give a series of parallel dark lines against a bright background (absorption spectrum),

Isaac Newton carried out much of the ground work on the refractive properties of light, but it was Fraunhofer, in the early 1800s who pioneered practical spectroscopy.

As an optician, he produced the first high quality glass prisms, and later, by combining these with a telescope, the first true spectroscope; the modern version of the prism spectroscope was developed in 1859.

Glass prisms are cumbersome and bulky, however, and Fraunhofer later went on to design the diffraction grating - a series of fine parallel lines engraved onto the surface of a mirror, which focused light directly without the need for a prism, and also had much higher dispersive powers.

He used this to study the spectra of the moon and planets, and later was the first person to study the spectrum of stars, such as Betelgeuse, and subsequently the dark absorption lines which bear his name, though it was a number of years before the true nature of these was fully understood,

In it's original form, Spectroscopy was an adjuvant to optical astronomy, and initially studied only the spectrum of visible light, but has now been extended to cover the whole range of the electromagnetic spectrum, and together with photography, has revolutionized our study of astronomical objects.

Spectra give a great deal of information about their source, both directly with respect to chemical composition, and indirectly about a range of physical properties, including temperature, mass, luminosity and distance.

They also give information about line-of-sight motion (towards or away from the observer) from Doppler displacement of the spectral lines, and perhaps spectroscopy's most notable contribution to cosmology was the expanding universe, based on Hubble's finding that the red shifts in galactic spectra were proportional both to their distance and to velocity of recession.

Spectroscopy was also central to the development of quantum mechanics, including Max Planck's explanation of blackbody radiation, Einstein's explanation of the photoelectric effect and Bohr's explanation of atomic structure and spectra.

Today it is also used in physical and analytical chemistry, because atoms and molecules have unique spectra, and these can be used to detect and quantify information about the particles themselves, and most research telescopes today are now fitted with spectrographs.

Spectroscopy was the first major advance in instrumentation since the telescope was invented, and allowed astronomy (which could only study surface features) to expand to become astrophysics (the study of physical and chemical properties) and that in turn made cosmology possible (the study of the whole universe) .

It has changed greatly since it's inception, however, and can now be applied to a variety of different types of radiant energy, including:

- Electromagnetic - radio, infrared, visible, ultraviolet, X-ray and gamma ray.
- Acoustic - involving radiated pressure waves.
- Mechanical - involving radiation energy waves in solid materials.
- de Broglie wavelength† - associated with the radiation energy of particles, e.g. electrons and neutrons, where the kinetic energy of the particle, determines its wavelength.

Astronomical Spectroscopy is theoretically applicable to any form of radiation, but depending on wavelength, other factors have to be taken into account with respect to accessibility.

---

† A quantum mechanical concept of wave-particle duality, where wavelength is inversely proportional to momentum.

Radio signals, for example, with much longer wavelengths than visible light, require large parabolic dishes receivers, while both relatively long infrared wavelengths, and very short UV and X-ray wavelengths are absorbed by (different) atmospheric gases, and therefore need to be carried out above the atmosphere, either by rockets or satellites.

High-quality reflection gratings were actually introduced for optical astronomy as early as 1900s, but these are limited by the size of the mirror, and many applications now use holograms combined with interferometry, to greatly increase resolution.

Radio interferometry was introduced shortly after WW2, but the real breakthrough was the introduction of aperture synthesis, combining signals from a collection of telescopes (optical or radio), to give an angular resolution equivalent to that of a telescope the size of the entire collection..

X-ray astronomy is now a branch of space science‡, carried out with space-based telescopes, such as the Einstein observatory and Chandra X-ray telescope, while Gamma-ray astronomy is perhaps the most recent addition, and the Fermi Gamma-ray Space Telescope was placed in low earth orbit in 2008.

Aside from its more conventional uses, spectroscopy also has a wide range of application within the solar system, where the light reflected from planets contains absorption bands due to the mineral content of exposed rocky surfaces, or the chemical constituents of atmospheric gases.

It is now also the major tool for studying exoplanets, once they have been identified, and around 1800 have now been discovered so for. These range from sub-stellar objects, so-called 'hot Jupiters', to Earth-like planets, many within the habitable zones of their parent stars, and spectroscopy has now identified compounds, such as alkali metals, and atmospheric gases, including water vapor, carbon monoxide, carbon dioxide, and methane.

12.6 Space Chemistry:

The precise chronology regarding the formation and evolution of the elements is unclear, but it is thought that only H and He, with possibly very small amounts of other elements up to Boron, were formed in Big

---

‡   Space science is now a discipline in it's own right, and studies issues related to space travel and space exploration (including space medicine) and a wide range of scientific and research projects carried out in outer space.

Bang nucleosynthesis (4.3). As a result, first generation (Population 111) stars were all metal-free[6], and also being composed solely of lighter elements, are thought to have been very large - up to 130x the mass of the sun.

We have no first hand knowledge of this era, however - only computer models and conjecture - as very large stars are always short lived, and no population 111 stars have ever been observed.

All metals (elements beyond helium in the periodic table) are produced entirely by stellar nucleosynthesis, and disseminated into space when a star dies, so that successive generations of stars become increasingly metal-rich.

Carbon and other organic elements thus build up in space, and the element of life are now actually common, not only throughout space itself, but also incorporated into every 'object' which interstellar dust and gas clouds condense to form. We have ample spectroscopic evidence to confirm this, and these elements in turn will eventually be incorporated into protoplanetary discs - and in the case of our own planet, progressed to form life.

Such a preponderance of 'elements of life' might suggest, on that basis alone, that life itself should be common, and especially organic life, if not other forms as well, and we discuss these possibilities in chapter 14.

Nevertheless, we still have no confirmation of the existence of life elsewhere, in any form, and if our obsession with 'intelligent life' seems over-optimistic, primitive life might be a more realistic objective, and our knowledge of interstellar chemistry, and molecular biochemistry, has increased rapidly over the past few decades.

Interstellar molecules are formed by chemical reactions within interstellar or circumstellar clouds of dust and gas, usually from the ionizing effects of cosmic rays, and are detected spectroscopically, mainly at longer wavelengths, in the radio, microwave or infrared regions of the spectrum.

Carbon-containing molecules in the interstellar medium were first detected back in 1937, and by the 1970s it had been established that interstellar dust consisted largely of complex organic molecules. Fred Hoyle and Chandra Wickramasinghe later demonstrating the existence of polycyclic aromatic hydrocarbons (PAHs) molecules in space, and suggested that these may have been important in the formation of early life on earth, NASA scientists have also shown that under interstellar medium conditions, PAHs can transform into more complex organic molecules - a step nearer to forming amino acids and nucleotides, the raw materials of proteins and DNA.

In June 2013, PAHs were detected in the upper atmosphere of Saturn's largest moon, Titan.

Quite a number of recent findings are also in keeping with the ideas of Panspermia (C9). In 2010, for example, fullerenes, which have been proposed as one of the possible candidates for 'seeding' life onto earth, was detected in the spectrum of a nebula, while the amino acid glycine, one of the fundamental building-block of life, was recently identified in material sampled from a comet.(9.4).

In 2011, scientists found that certain organic matter, whose structure was so complex that it had previously been thought to arise only from living organisms, could be created naturally in stars[7], and this suggested that organic compounds, introduced on earth by interstellar dust particles, might serve as basic ingredients for life.

In August 2012, astronomers detection molecules of the sugar glycolaldehyde, used in the synthesis of RNA, in a distant star system, suggesting that complex organic molecules might form along with a star itself, prior to the formation of planets, and then migrate to planets at a later stage.

Quantum tunneling has also now been suggested as an explanation for certain reactions which were taking place more rapidly than might be expected, in spite of intramolecular energy barriers which they would not normally have been expected to overcome.

Overall, Sagittarius, near the center of the Milky Way galaxy, is a particularly rich region for detecting interstellar molecules, with one giant molecular cloud in which up to 50 compounds have now been identified[8].

12.7 Surface Exploration:

We know a great deal about the physical and chemical composition and environmental conditions of all of the major planets, both as they are now, and in certain cases, also with respect to their geological past, and have covered these aspects in chapter 5.

In addition, an increasing amount of information has been obtained about the surface features and/or atmospheric conditions of the moon and other satellites, larger asteroids, and a number of comets.

The visible surface features of most of these objects have been studied and mapped with earth based instruments over the years, but the information we are now interested in - geophysical and chemical properties, with respect to possible suitability for the presence of life - has been obtained only over the past few decades, by successive generations of space probes, either placed in orbit, 'flybys', 'landers' and most recently 'rovers' and 'mini- laboratories'.

Detailed photography has provided a pictorial record of many features we previously knew nothing about, while on-board instruments, spectroscopes and radar, have provided detailed first- hand information of atmospheric and surface features - geochemical, geophysical, surface terrain and sub-surface information, both by probes and by radar, while on Mars, rovers are currently searching for microfossils or even microbial life itself (5.11).

The lunar surface has been explored more thoroughly than any other region of the solar system, by all of these means, and in addition to 6 manned landings - a total of 12 days spent on the moon's surface, including 79 hours of astronaut EVAs - samples of rock and surface material returned to earth. A Single microorganism, found in one of these samples, was subsequently confirmed as terrestrial contamination.

Of the other satellites in the solar system, Phobos, the larger of the two Martian moons, has been explored the most thoroughly, perhaps because it's unusual orbital characteristics led to a proposal, by a Russian astrophysicist in 1958, that Phobos, which is roughly oval in shape and measures about 16 x 12 miles, was in fact a hollow metal sphere of artificial origin.

Multiple missions, both to Mars and to Phobos itself, have long since disproved that, and the Viking missions in the 70s confirmed Phobos as a 'rubble pile' of porous rather than solid rock. It's heavily cratered surface has been recorded in detail, but attempt to obtain samples have all failed.

The 4 Galilean moons of Jupiter, of which Ganymede is he largest moon in the solar system, have all been explored in similar ways, Titan, the largest of Saturn's moons, is the only natural satellite known to have a dense atmosphere, and the only object, other than Earth itself, with stable bodies of surface liquid, which have now been shown to include lakes, seas, rivers and even rain - clear evidence of a hydrological cycle[9].

The Cassini probe in 2004 carried out detailed high resolution mapping of surface features, both photographic and by radar, while the

Huygens probe, which landed successfully in January 2005, found rocks composed of water ice, areas covered with water and hydrocarbon ice, and rocks showing characteristic evidence of erosion.

A number of asteroid flybys have also been made, and The 'NEAR Shoemaker' space craft made a year long study of the near- earth asteroid Eros, and finally terminated with a surface landing. It carried the most comprehensive equipment of any interplanetary mission, including X-ray and gamma ray spectrometers, a near-infrared imaging spectrograph, a multi-spectral camera fitted with a CCD imaging detector, a laser rangefinder, and a magnetometer.

Data from this mission was used for a number of purposes, including the study of magnetic fields, the solar wind and the relationships between asteroid, meteorites and comets Asteroids are also of interest for two other reasons - the potential dangers of asteroid impact (7.11) where an object more that a few miles in diameter could threaten another extinction event; and the possibility of mining, where asteroids or spent comets could provide a valuable source of minerals and volatiles.

Comets have also been the target of a number of flyby missions, including one impact landing and one sample and return mission in 2006, and together with meteorites, may have played a role in the dissemination of life.

The Allan Hills meteorite (9.4), however, also exemplifies the difficulties of dealing with potential 'evidence' of this sort, and typifies the general problem that if different interpretations are possible, unanimity becomes ever less likely.

All of the above missions were exploratory, and none have so far shed any light on the possibilities of life existing elsewhere than on our own home planet; but they all have a dual objective, nonetheless, beyond the tasks they were designed to carry out, and that may sometimes be as useful as their primary objectives - research and development, of skills and technology to match ever more demanding missions and objectives of the future.

Other stars and other planetary systems may seem little more than 'Kepler data' at the present time - but they are the reality of the future, and the only uncertainty is 'when?'.

12.8 ExoMars (Exobiology of Mars):

This is an advanced astrobiology project, jointly between the European Space Agency (ESA) and Russian Space Agency, to search for evidence of Life, past or present, on Mars.

The project has multiple funding, of which the UK is the second largest contributor, but NASA has had to withdraw, for financial reasons, due to the escalating costs of their own Next Generation Space Telescope, the successor to Hubble.

The ExoMars project includes a variety of equipment and instruments, to be delivered to Mars in two installments[10].

The Trace Gas Orbiter (TGO) is a telecommunications orbiter and atmospheric gas analyzer, due to be launched in January 2016. It is scheduled to go into Martian orbit in October 2016, when it will deliver a stationery 'lander' (known as Schiaparelli) onto the surface, and then go on to map methane and other atmospheric gases.

Schiaparelli, in turn, will carry out a study of surface environmental conditions, including wind speed and direction, humidity, pressure and surface temperature, and the influence of electric fields on the dust storms which are such a feature of the Martian environment.

These findings, together with atmospheric information from the orbiting TGO vehicle, will then be used to decide a landing site for the second part of the project - an ESA built Martian 'rover' - due to be delivered by a Russian heavy lift Proton launch vehicle in 2018, and scheduled for landing in January 2019.

The 'rover' is built to navigate automatically over the Martian surface, and comprises an exobiology laboratory, known as the 'Pasteur analytical laboratory', and which includes a 2-metre (6.6 ft) drill for obtaining deep-surface samples.

The mission has a very wide remit, including looking for signs of biomolecules and biosignatures, water and geochemicals, to understand the evolution and habitability of Mars, and to evaluate conditions and hazzards preparatory to a future manned landing.

12.9 Kepler Space Observatory:

The Kepler telescope is part of the NASA 'Discovery Program' - a series of dedicated low-cost missions, to explore the solar system.

The instrument itself is a space-based reflecting telescope[11], designed specifically to survey local regions of the Milky Way for extra solar planets, and in particular, earth-sized planets within the habitable zone (5.10) of their parent star.

Kepler was launched from Cape Canaveral on March 7th. 2009, into to a heliocentric earth-trailing orbit, with a period of 372 days. It's initial mission was planned for 3.5 years, but a number of unforseen difficulties meant that more time was necessary to accomplish objectives, and the mission was extended, provisionally to 2016. Subsequently, however, it was changed to a pre-arranged backup (K2) mission, with different technical requirements, and which may last for a further 4 years.

The Kepler spacecraft has a basic cylindrical design, measuring 15ft x 9ft, with a single 4.6 ft mirror - the largest single mirror located outside Earth orbit. It's sole instrument is a photometer, continuously monitoring 145,000 main sequence stars in it's local area[15]. All data is transmitted to earth, and analyzed to detect changes in brightness resulting from planets crossing the disc of their parent star.

The Kepler camera consists of 42 high-resolution CCDs, with a fixed field of view, which covers 100 square degrees, or about 0.25% of the whole sky.

Data is read out every 6 seconds, but 95 million pixels is more than can be stored on board, and therefore only pre-selected pixels, amounting to about 6% for each star of interest, are collected for analysis. Nevertheless, for earth-sized planets the signal level is so close to the noise level, that many more transits have to be recorded than was originally planned, to confirm identification.

The Kepler's original mission had a number of objectives, including:

- To determine numbers of earth-sized planets in habitable zones.
- To determine their orbital parameters and value 'ranges'.
- To estimate the numbers of planets in multiple star systems.
- To determine the properties of stars which form planets.
- The K2 mission included study of supernovae, star formation and other solar system bodies such as asteroids and comets.

A minimum of three transits (among other criteria) was required for confirmation of an exoplanet, while 7,500 stars were subsequently found to be 'variables', and had to be excluded from the original list.

As of 12 September 2014, 1,822 exoplanets have been discovered, in 1,137 planetary systems, including 467 multiple planetary systems[12].

Kepler has also detected a few thousand unconfirmed 'candidate' planets, and about 11% of these may be false positives.

12.10 Exoplanets (extra solar planets):

These are defined as planets which do not orbit the Sun, but instead orbits a different star, stellar remnant, or brown dwarf.

There is nothing new about the concept of 'other planets', and over the ages, philosophers and scientists alike have always accepted that such objects existed somewhere, though not in any way that could interfere with man's existence on this planet .

In the nineteenth century, claims were even made to have detected such objects, even though instruments and telescopes available at that time, could not possibly have accomplished it.

The first confirmed detection of an exoplanet was in 1992, with the discovery of several earth-mass planets orbiting a pulsar[11], while the first confirmation of an exoplanet orbiting a main-sequence star (i.e. similar to our own sun) was in 1995, when a giant planet was detected in a four-day orbit around the nearby star 51 Pegas[12].

For the next decade, the discovery of exoplanets was slow (by present standards), but this changed dramatically once Kepler was launched. The detection rate has escalated ever since, and is now the main field of study for the new discipline of astrobiology.

This pattern of progress and change is becoming increasingly common, in all areas of science and technology, and progress in detection of exoplanets is a good example of the biological equivalent of Moore's law:

Figure 12.1
Annual Detection of exoplanets, 2000 - 2014:
(K Kepler launched March 2009)

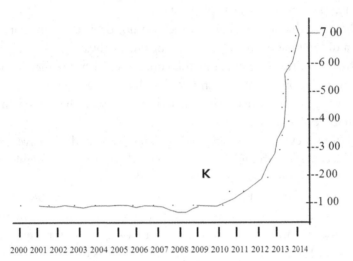

On average, every star has at least one planet, while approximately 1 in 5 Sun-like stars have an 'Earth-sized' planet orbiting within the habitable zone; the nearest of these is probably within 12 light-years distance from the Earth.

Hence, if we assume 200 billion stars in the Milky Way, that would be 11 billion potentially habitable Earths, rising to 40 billion if red dwarf stars are included, while there could be literally trillions of unattached, free-floating planets, just in this galaxy alone.

The overall proportion of stars with planets is uncertain, however, because not all planets can yet be detected, and the radial-velocity and transit methods (which between them are responsible for the vast majority of detections) are most sensitive to large planets in small orbits, while the Kepler observatory, using different methods, preferentially detects smaller planets.

A number of free floating planets, which do not orbit any particular star, have also been identified, and if these are gas giants, they are often regarded as low-mass brown dwarfs.

Thus, there is a preponderance of larger, 'hot Jupiters' planets over smaller planets, and a 2005 survey of radial-velocity-detected planets, found more than double the frequency of hot jupiters planets compared with those detected by the Kepler spacecraft.

This imbalance will change over time, however, because Kepler has a much higher detection rate than the older methods, but at present it is not possible to say what the true proportions are[11][12].

With increasing confirmation of exoplanets, the existence of extraterrestrial life is now regarded as a virtual certainty, and accepted as such by the scientific establishment, with on-going projects such as SETI, and the relatively new discipline of astrobiology, to study the biophysiology of life, and the potentials for human exploration of space, and manned missions to other planets.

The IAU system of definitions can sometimes be a little confusing, and the standard definition of a 'Planet'(5.4) does not cover exoplanets. However, they issued a working definition more recently, which was modified in 2003, and contained the following guide lines[13]:

- Objects with true masses below the limiting mass for thermonuclear fusion of deuterium (currently calculated to be 13 Jupiter masses for objects of solar metallicity) that orbit stars or stellar remnants are 'planets' (no matter how they formed). The minimum mass/size required for an extra- solar object to be considered a planet should be the same as that used in our solar system.

- Substellar objects with true masses above the limiting mass for thermonuclear fusion of deuterium are 'brown dwarfs', no matter how they formed or where they are located.

- Free-floating objects in young star clusters with masses below the mass limit for thermonuclear fusion of deuterium are not 'planets', but are 'sub-brown dwarfs' (or whatever name is most appropriate).

The difficulties in directly detecting exoplanets have little to do with instruments and equipment, but rather the fact that any stars will be at least a million times brighter than the faint reflected light from any planet, which is therefore swamped to the point of invisibility.

There are techniques, however, in which a coronograph can be used to block-off light from the star, and a small number of very large planets, orbiting well out from their parent star, have been viewed directly in this way.

Alternatively, many stars have a brighter infrared image (especially the H-alpha band†) than at visible wavelengths, and the Gemini South Telescope in Chile, for example, which saw 'first light' in November 2013, is designed to directly detect young gas giants in this way. It will only have limited availability, however, and whether it offers a useful alternative or not, remains to be seen.

The vast majority of exioplanets have usually been detected by the Transit Method, which identifies the tiny changes in brightness of the parent star, as the planet passes in front of it's disc. It depends very much on the size of the planet, however, and tends to have a high incidence of false positive 'confirmations'.

Nevertheless, for suitable planets, this method allows their radius to be measured, and in some instances, the planet's atmosphere can also be examined spectroscopically.

It's main disadvantage, is that it is only applicable to planets crossing the direct line-of-sight between a star and the observer, whereas random planetary systems may present at any orientation If a planet orbits multiple stars, however, or the planet has moons, its transit times can vary significantly, and although no new planets have been discovered with this method, it has been used successfully to confirm planets which have.

Conversely, if more than one planet is present, each will slightly perturb the orbit of the other, and the variation in transit times which result, can be a pointer to the existence of another planet, even if that planet is not itself transiting at the time. Again, few planets have been discovered in this way, but numerous planets have been confirmed.

An orbiting planet can also be detected by the Doppler displacement of spectral lines, due to the Radial velocity changes as the planet either approaches or recedes from the observer. This method has the big advantage of being compatible with a wide range of stellar characteristics; however, to determine the planet's mass, the radial velocity of the parent star has to be identified and allowed for, otherwise it is only possible to set a lower limit for mass.

For a long time, this was the most successful method for detecting exoplanets, but has now been superceded by the capabilities of the Kepler

---

† A specific deep-red visible spectral line in the Balmer series created by hydrogen with a wavelength of 656.28 nm

space observatory, and since that became operational, in March 2009, the detection rate of extrasolar planets has progressively escalated (Figure 12.1).

Gravitational microlensing relies on the relativistic property that the path of light, passing close to a massive object, is bent, in effect acting like a lens, to magnify the light coming from some more distant background star. Planets can be detected in this way, through changes in the level of magnification as they circles around the star.

Astrometry is more theoretical than practical, and involves recording precise serial measurements of a stars position, where very small changes due to the presence of a planet might be detectable. As with some of the others methods, however, very few planets have actually been detected in this way, though many known planets have been successfully confirmed.

Pulsar timing relies on the fact that pulsars (ultra dense remnants of a collapsed star) emit extremely regular radio signals as they rotate, and any planet orbiting one of these will distort the regularity of these signals. Just how common planets orbiting pulsars are, is not known, but the very first extrasolar planet was actually discovered in this way, in 1992, and at least 5 planets, associated with 3 different pulsars, have been discovered since.

A few variable stars have very precise frequencies too, and a number of planets related to these have also been discovered in this way.

There are a number of other possibilities by which planets might reveal themselves, and these include: Reflection modulation, due to a planet displaying phases as seen from earth; Relativistic beaming, where the light from a star varies with the proximity of a planet; Ellipsoid variations, where very massive planets can slightly deform their parent star, with corresponding changes in it's brightness; Polarimetry, which relies on being able to distinguish the polarized light reflected off a planet's surface, from the unpolarized light being emitted by it's star; and Circumstellar discs of space dust, which surround many stars, and can sometimes show spectral changes, which could imply the presence of a planet.

References (Chapter 12) :

1. P. Lowell, Mars (1895); Mars and Its Canals (1906); Mars As the Abode of Life (1908).
2. M. Kidger, Astronomical Enigmas: Life on Mars, the Star of Bethlehem, and Other Milky Way Mysteries (2005).
3. M. J. Seifer, Martian Fever (1895–1896) Wizard : the life and times of Nikola Tesla: biography of a genius (Secaucus, New Jersey, 1996).
4. K. L. Corum, J. F. Corum, Nikola Tesla and the electrical signals of planetary origin (1996).
5. Wikipedia article (1 January 2014) Search for extraterrestrial intelligence (The SETI League and Project Argus).
6. A. Heger, S. E. Woosley, The Nucleosynthetic Signature of Population III. (Astrophysical Journal 567 (1): 532–543).
7. D. Chow, Discovery: Cosmic Dust Contains Organic Matter from Stars (Space.com, 26 October 2011).
8. J. B. Kaler, The hundred greatest stars, Copernicus Series (Springer, 2003).
9. Space.com (6 July 2014).
10. Wikipedia article (4 March 2014), ExoMars.
11. Wikipedia article (5 March 2014), Kepler (spacecraft).
12. Wikipedia article (12 September 2014), Exoplanets.
13. Working Group on Extrasolar Planets: Definition of a 'Planet' (IAU position statement, 28 February 2003).

Chapter 13

INTERSTELLAR TRAVEL

"If they won't come to us, we must go to them"
- Enrico Fermi

Interstellar travel has been an integral part of science fiction for so long, that not only has it lost much of it's mystique, but through familiarity, has almost come to be accepted as science fact. It would certainly be essential to any form of intelligent extraterrestrial life, however, and therefore given that we have now come to accept the reality of such beings, it can only be a matter of time before interstellar travel too becomes a necessity, if we are ever to retain our role as the dominant species.

That is no doubt how the general public would see it, but for the scientists who have to bring it about, the future is much less certain. Only they are in a position to extrapolate from present knowledge, and the reality is, that interstellar travel is not within the scope of existing science, either now or in the foreseeable future, while there are even some who do not believe it will ever happen.

There are too many practical difficulties, not least - though one that is rarely mentioned - selecting an arrival point, navigating during the journey, and the formalities of arrival at some arbitrary point divorced from necessary facilities and technology. None of these things can be foreseen and allowed for in advance, quite apart from the difficulties of the journey itself.

Proxima Centuri, the nearest star to the sun, is 4½ light years away (about 26 trillion miles), and there is all the difference in the world between exploring distant regions on earth, perhaps over a few thousand miles, and explore the infinite voids of interstellar space.

Indeed, it is ironic that we should even be thinking in terms of such distances, when there are regions on earth, just a few miles beneath the surface of the oceans, which we still know nothing about, because we do not yet possess the necessary technology to explore them.

These anomalies aside, however, and it is also a mistake to assume that interstellar travel would necessarily resolve the question of alien life, in the unlikely event that it ever did become possible before that question had been settled, one way or the other.

The impracticality of 'arrival' will always stand in the way of completing any successful journey, while there are many other factors en route mitigating against traveling by spaceship at all.

The physical and psychologically stresses of prolonged confinement and uncertainty, are things we know very little about, and would also be difficult to simulate, but could significantly affect the outcome of such a missions nonetheless.

Interplanetary travel will no doubt be routine by then, and there are certainly ways in which that could be used to simulate certain of the difficulties encountered, but it is the duration of an interstellar journey where the real difficulties lie, and unless we can reduce these, at least in terms of perception, if not reality, it will neither be practical or possible for man to explore the universe in any physical sense. The reality is, however, that we can explore the universe more easily and in greater detail, from earth, using the space based instruments we have, together with new technology as and when it becomes available.

Concepts of distance can be expressed in a number of ways, but for interstellar purposes, it is usual to do this by equating distance with time, using the speed of light - 186,000 mps (which is constant in the vacuum of space, and also cannot be exceeded by any material object) - as a yardstick.

One light years (ly) - the distance (in round numbers) a beam of light would travel in that time - is 6 trillion miles or 10 trillion kilometers. Such numbers are impossible to visualize, but scaled down and applied to the solar system, and they become much more meaningful.

Thus, the distance from earth to moon is a mere 1.3 seconds; the distances between the nearest and furthest planets in the solar system range from 3 minutes to 4 light hours, and by comparison, the distance of the nearest star is just over 4 light years.

There are 54 stars within 16 light years of earth, the majority being red dwarfs, and some of these are known to possess planets (12.10). The Milk Way galaxy (5.3) by comparison, has a diameter of 120,000 ly, and the solar system lies about 25,000 ly from the galactic center,

Voyager 1 (5.11), the fastest outward bound spacecraft, is traveling at 1/18,000th the speed of light (10mps or 36,000 mph). So far, however, it has only covered 1/600th of a light-year since it was launched 30 years ago, and will take 80,000 years to reach the nearest star.

Nevertheless, such speeds would be little use for interstellar travel, and the technology for that has not yet even been envisaged. It will be totally different, however, both in concept and design, from anything which exists today, and is likely to embrace some or all of the following principles:

- No form of conventional energy sources, e.g. as currently used for launching interplanetary probes and satellite missions, will be suitable for interstellar travel.
- To travers interstellar distance in 'practical intervals of time', would have to involve one or other of the following:

    - Faster than light travel (FTL - 13.6)
    - Sub-light velocities
    - Time dilatation (13.3)
    - Hibernation
    - Life extension (13.9)

- All such journeys will involve distortions of time (and/or of space) in some form or other.
- Primary Time Travel, by itself, may prove to be the most practical method, but it will also form part of the rationale behind a number of other possible methods (13.6), including:

  - Closed timelike Curves
  - Wormholes
  - The Alcubierre Drive
  - Tipler's Cylinder

- Other physical properties (13.6), which we still know very little about, and for that reason cannot exclude unforseen potential uses, might incorporate:

  - Entanglement and non-locality
  - Cosmic Strings

Hence, for practical travel purposes over interstellar distances, Time Travel will be an essential part of the principle involved, irrespective of the primary mechanism.

Eventually, Time Travel on it's own might come to replace all of these, as the primary way in which human life could explore it's parent galaxy, and in the very long term, the only way in which it would be possible to explore the wider universe at an inter-galactic level.

Hence, it is convenient to look at time travel itself, including theoretical background and the physical principles involved, as an umbrella topic in its own right, and a necessary preliminary to understand all of the other potential methods of travel, which will be introduced and explained in their proper context, later in the discussion.

13.1 Time Travel:

Time travel, as conventionally envisaged, is the concept of moving through time, while the physical environment, although changing appropriately during the journey, retains it's initial position in space, and is unchanged when the traveler comes to return.

The idea of moving through time is not new, and indirect references to it date back to the early folklore and mythology of many cultures. The ghosts of Ebenezer Scrooge, in Dicken's Christmas Carol, published in 1843, who visit both past and future, are a relatively modern example of this sort. However, the first formal account of time travel is often attributed to the short story

'The Clock That Went Backward'[1], by Edward Mitchell, published in September 1881, in which 3 men travel back in time specifically to change the outcome of a decisive battle.

This pre-dates H. G. Well's classic 'Time Machine' by 15 years, and also differed from that in two important ways. Wells describes a journey into the future, primarily as an observer, and though his traveler interacts with other beings, he changes nothing; Mitchell's travelers, by contrast, do so for a purpose - still one of the most contentious issues even today - to change the future by altering the past; in their case successfully.

Time travel bears no relation to conventional 'movement' between two points, which depends on such physical factors as the nature of the path, or the presence of intervening objects. Light, for example, can be slowed down or the signal altered entirely, by passing through dust and gas clouds (extinction effects), or it's path bent by gravitation, from passing close to some massive object, such as a galaxy (gravitational lensing).

Time travel is a different concept of 'movement' altogether, and can best be thought of in terms of 'relocation' within the framework of time, which alters physical location, but is independent of intervening physical change. In that sense, time travel will be instantaneous, irrespective of the size of physical separation, and likewise, as there is no movement in space, it does not violate the cosmic speed limit.

A traveler would also expect that the place he started out from would still be there when he returned, irrespective of when the return journey took place. Yet that may not be so; once he becomes detached (in time) from any particular starting point, that will nevertheless continue it's normal movement through space, due to the earth's rotation and orbital motion, and there is no logical reason why it should still match up with some arbitrary time in the future, so that the traveler could in fact end up somewhere quite different, or even in surrounding empty space.

13.2 Einstein's universe:

Time travel, as commonly envisaged, concerns changing from one period in time to another, in relation to a fixed background of

3-dimensional space, which is both easy and convenient to visualize, but bears little relation to the real thing.

separate; three dimensions is how we normally think of our environment, perhaps because our awareness of 'space' is based on familiarity with objects that we can see and touch, while there is no such equivalent 'contact' with time.

Newton's classical universe was based on the concept of space and time as two entirely separate, immutable properties of nature, and it lasted for over 200 years, because it was both convenient and appropriate to the needs of the real world at that time.

It was Einstein, however, at the turn of the century, who revolutionized cosmology with his theory of Special Relativity, which defined 'time' as a separate (4th) dimension - analogous to (but different from) the other 3 space dimensions - and also introduced the concept of spacetime, in which 'space' and 'time' were no longer the separate properties envisaged by Newton, but rather united together into a single entity.

Einstein's universes[2] was now seen as an infinitely large volume of expanding spacetime, finite but unbounded, and with a range of physical properties very different from those of Newton's universe.

It is a difficult concept to grasp, based on tensor equations, rather than something which can be readily visualized - there is no boundary, for example, in any conventional sense, and the universe is not expanding 'into' anything.

The main differences, however, are in the geometry. By contrast with flat Euclidian space, where straight lines unite single points into continuity, and define the shortest distance between them, the geometry of 4-dimensional spacetime is determined by the quantity of matter it contains; spacetime is curved; the path of light is bent, to an extent related to the proximity of matter (mass) at any given point; geodesics are now the maximum space-time interval between events, and moving objects travel along wold lines, which unite events (in time), as opposed to points (in Euclidian space).

It is only within such a framework, where time is integrated into the fabric of the universe, that the concept of moving through time has any meaning, and time travel becomes a theoretical possibility. What really matters, however, is whether it would be possible in practice.

13.3 Theoretical Background (1.10):

Spacetimes is a continuous environment in which all physical events take place. The basic elements of spacetime are events, which occupy a specific position in space, at a specific points in time; although they exist in four dimensions, they can be illustrated on a 2-dimensional space-time diagram, in which 'time' is plotted against one dimension in 'space' only. The location of any event is then specified by two coordinates, t and x, corresponding to the 'time' and 'space' axes of the graph[2].

The history of an event, i.e. changes in location over a given period of time, is then shown by a line on the graph, joining up points plotted for each different location (t, x) - the world line for that event. For convenience, any physical object, such as the earth, or even a beam of light, is regarded as a single point, so that how it behaves over any extended interval (i.e. it's history) can always be depicted as a world line in this way.

In flat Euclidian space, points are compared in terms of their distance apart. In spacetime, however, where events are separated both in space and in time, space-time intervals are used instead. These have complex properties, but two of the most important are: that they are the same for all observers, no matter 'where' or 'when' an observer is situated (i.e. invariant under Lorentz transformation); and that they divide the universe into three different categories, depending on whether causal contact is possible between them or not.

Causality depends on the temporal relation between past and future, and can be described as follows:

> Things which exist in the present depend on their predecessors (i.e. things in the past), and at the same time, are themselves antecedents for events which have yet to happen (i.e. the future).

In other words - everything which exists in the present today stems from the past, and determines the future. This conforms with the unidirectional nature of time, and clearly, if we can override that sequence, by moving within time, a great many anomalies could arise (13.7).

Most theoretical work on time travel is based on Einstein's equations[3], from the concepts of variable time and time dilatation, in special relativity, to general relativity, which extended the special theory to include gravitation, and the final field equations which have given rise to a wide variety of complex interpretations, with respect to the properties and implications of 4-dimensional space- time geometry.

From a practical standpoint, traveling forward in time is theoretically possible (Special Relativity). The return journey, however, traveling backward through time to the present, has not yet been shown to be possible, though neither has it been shown not to be possible.

The two journeys in many respects are completely different, and though it may seem logical to regard the return journey as an 'integral part of a round trip', there are a number of difficulties which will remain problematic until we can actually test them out.

The arrow of time, for example, moves in one direction only - forward, based on correlation with entropy, which can never decrease. Time cannot flow backwards, and while it is possible to travel into the future, so far as we know (see below) it is not possible to travel into the past, i.e. against the flow of time.

Hence, having moved forward into the future, a traveler wishing to return to the present, would have to travel backwards in time in order to do so, and that may not be possible. This problem (to the best of the author's knowledge) has not been addressed before, and hence we make the following prediction:

> If time travel ever becomes possible, it will be in one direction
> only - forward, and hence no successful traveler will ever be able
> to return to his own time.

There is always the possibility, however, that nature will make a distinction between moving 'to and fro', and traveling back in time de novo from our natural 'present', though otherwise, that could be one reason why we have not been visited by travelers from the future.

The concept of moving in time, as a practical reality, stems indirectly from Einstein's study of simultaneity[2], in which he showed that time measured in any two moving environments, had different properties, depending on the velocity difference between the two environments; and specifically, time slows down as the velocity of the environment in which it is measured increases (time dilatation[2]).

Einstein later showed that several other properties of nature were similarly affected by movement, and adapted the Lorentz transformation equations to describe these changes. The equation for time has the following form:

In equation
$$Tm = Ts\sqrt{1 - \frac{v^2}{c^2}}$$

where Tm = time measured on a moving clock
  Ts = time measured on a stationary clock
  v = speed of the moving clock
  c = velocity of light.

Hence, time measured on a moving clock (Tm), slows down as velocity v increases, compared with time measured on a clock at rest (Ts), and would stop altogether if v could ever reach the velocity of light (c). The rate at which time passes depends only on the instantaneous velocity of the clock, and is not affected by acceleration (clock hypothesis).

13.4 Terminology and Interpretation:

Terms such as 'moving' and 'traveling' are ambiguous; they imply changes in spatial position only, and are better avoided. Professionals prefer to discuss time travel in terms of closed timelike curves, which are world lines that double back on themselves, to form a closed loops, such that any object traversing spacetime along one of these will arrive back where it started.

These rather unusual paths are nonetheless 'allowed' by certain solutions to Einstein's equations, and can therefore be interpreted in ways which would be consistent with time travel. Quite a number of uncommon physical properties are 'allowed' in this way, by theories such as relativity and quantum mechanics, for example, though not actually predicted; while others, by contrast, can be specifically forbidden

Nevertheless, 'allowed' is such a vague term, that a variety of interpretations may seem equally valid; what is important, however, is that being 'allowed' has nothing to do with whether something actually takes place or not, and statements (not uncommon) in which the context might suggest otherwise, should be treated with reservation; the reverse of course must be true.

13.5 Realties of Time Travel:

There are certain phenomena in physics which at first sight may suggest exceptions to the ordering of events, and in that respect technically could be regarded as examples of time travel. Some of these seem to involve faster than light movement (e.g. Cherenkov radiation[4]), while others seem almost instantaneous (e.g. quantum entanglement). Tachyons[7], on the other hand, which would normally travel faster than light, are not permitted in current versions of string theory[2]. In so far as none of these would allow transfer of information, however, they are not regarded as breeching causality, and the reality is that no verifiable examples of time travel in the truly accepted sense, have ever been demonstrated.

Time dilatation is ambiguous, because 'compression' of time is not the same as 'traversing' time, and many would not regard 'compression' as 'time travel' in the normally accepted sense.

It is a well recognized phenomenon in physics, none the less, and now has to be allowed for in much of the precision technology which exists today.

The rate of decay of fundamental particles, for example, is clearly affected by movement; time measured in earth orbit on atomic clocks, passes more slowly than on identical clocks on the surface; clocks belonging to the International Global Positioning network have to allow for these minuscule differences, and even on the international space station, orbiting at almost 5 miles/second, over a 6 month period, an astronaut would age 0.007 seconds less than his colleagues back on earth[5].

Some of these effects are offset by gravitation, which also causes time dilatation. The stronger the gravitational field, i.e. the nearer the clock is to the source of gravitation (lower gravitational potential) the slower it runs. As with velocity, this effect only becomes significant under extreme conditions, for example in the vicinity of a black hole. It still has to be taken into account for satellites in earth orbit, however, where the effect is opposite to that due to velocity, because a clock in orbit is further from the source of gravitation, and therefore will run faster, than a similar clock on the earth's surface.

The twins paradox[6], although based on dilatation, is nevertheless widely used as a 'genuine' example of time travel, which at one stage almost acquired academic status, with numerous papers in top scientific journals.

One of a pair of twins makes a round trip journey to a nearby star system, and apart from periods of acceleration at the start and finish of the journey, travels at just below the speed of light throughout. Since time passes more slowly under these conditions, due to time dilatation, than on clocks at home, the traveling twin ages very little during the journey, but on arriving back, still a young woman, finds her sister elderly and frail.

Since all time phenomena are interchangeable, the paradox is not so much the difference in ageing, but rather why we cannot use reciprocity to reverse the role of the two sisters, with earth receding from the spaceship at close to the speed of light, clocks on earth running slow, and the sister 'at home' the one who remains young?

The answer in this case, is that to be an observer (i.e. a person who 'observes' a time phenomenon, but is not affected by it), a persons must remain in the same inertial frame, in this case, for the duration of the journey. Since this only applies to the sister on earth (The traveling sister underwent two episodes of acceleration / deceleration) only her clocks will run normally, and therefore she will be the twin who is unaffected, and will age in the normal way.

This paradox, nevertheless, has attractive potentials for interstellar, or even intergalactic travel. Such travelers are likely to be in the 20-60 year age range, and therefore one might conclude that even traveling at close to the speed of light, they could never travel further than about 40 light-years from earth, and there are very few star systems within that distance.

There is another way of looking at it however. If the spaceship initially accelerates at just 1g, which will maintain normal sensation of weight during this stage of the journey, it will only take 354 days to reach a fixed cruising speed of just a little under that of light[6]. In these circumstance, time dilatation will increase their physiological lifetime to thousands of years, though their subjective perception of time will not change, and the space ship can traverse huge distances, embracing large numbers of potential star systems, while the travelers remain young.

Unfortunately, however, that scenario too would be impossible, for a number of reasons, not the least being that if they did decide to stop and explore another star system, the benefits of dilatation would be reversed as the spaceship slowed down, and they would no longer be young on arrival; but in fact they never would arrive, because as the spaceship started to accelerate at the very outset, their weight would increase progressively (with velocity), and long before reaching cruising speed they would be crushed to death.

Aside from time dilatation, however, there are no other physical procedures or properties that could move an object forward in time.

Backward time travel is, nevertheless, 'allowed' by Einstein's equations, and there are a variety of theoretical ways in which this might be accomplished (13.6), although all are too complex to be possible with present day technology.

Nevertheless, so long as time travel is considered an achievable objective, and irrespective of direction, too much attention is still being give to 'getting there', and too little (or none at all) to 'coming back'. The reality is, that none of the method now being discussed (13.6), including time dilatation, can be used both ways, and that any successful journey can only end in a time frame which could not provide facilities for getting back - with the possible exception of advanced technology in a suitably 'distant future'.

13.6 Physics of Time Travel (C1.10):

There are several ways in which certain physical properties might be used to circumvent the sequential nature of time[2][5]:

(1) Speed of light (FTL). Maxwell's equations[6] showed that the speed of light (c) was constant, and must be the same for all observers (whether they themselves are moving or not); while the equations of special relativity imposed the speed of light as an absolute limit for any form of movement within the universe (cosmic speed limit).

The relativity of simultaneity[2] allows, in certain circumstances, for time to pass at different rates in different reference frames, and therefore observers within these may disagree about the timing, or even the sequence, in which a given event may occur. If a signal passes between these two frame, however, so long as it does so at the speed of light, both observers will agree, as would be expected, that it set out before it arrived.

If such a signal could travel faster than light, however, this might no longer be true, and there will always be a situation in which one observer will claim to see the signal 'before' it was actually emitted. That would be a direct reversal of causality, and could therefore be interpreted as 'time travel'.

Nevertheless, the principle is sound; all events are interpreted in relation to the time frames in which they occur, or are perceived, and the cosmic speed limit relates different frames in such a way, that correct causal sequences are always preserved. This would no longer be true, however, if information could be transferred at speeds greater than that of light.

The practical criteria for FTL travel, however, is the transfer of information, and while there are a number of theoretical situations in which individual photons might exceed the speed of light†, there are no known circumstances in which information can be exchanged at greater than the speed of light.

Nevertheless, there are two hypothetical proposals for actual faster than light travel, and while these are not consistent with known laws of physics at the present time, there are 'grey areas' on the horizon, and it may be that some as-yet-to-be-discovered properties of matter could make them possible in the future; or there could be advanced aliens who are using them already.

It would involve concepts of negative mass, however, and propulsion would be by contracting space in front of it, and expanding space behind, so that vast distances could be traversed in very little time, and the net effect would be equivalent to traveling faster than light, without actually doing so.

An analogy which might help to illustrate these changes, would be that of a long train, traveling between San Francisco and New York, which drops off it's rear carriage at each stop along the way; the center of the ever shrinking train therefore moves forward at each stop, and in this way, the center of the train is traveling forward more quickly than the train itself!

This propulsion principle is analogous to that of the fictional Warp drive, used in Star Trek, which derives it's energy from matter/anti-matter annihilation.

(2) Wormholes are another hypothetical possibility for faster than light travel, permitted under the field equations of general relativity, which also

---

† In the Casimir vacuum, between two perfectly smooth metal plates, one micrometer apart, vacuum energy is reduced, and theoretically photons will cross this gap at a speed greater than c. The excess speed would be minute, however, and it is doubtful if this effect has been observed in practice.

involves 'warping' of spacetime, and have been proposed as a means of time travel[5][8].

Only certain wormholes are suitable, however, and designated 'traversable'. These were first introduced by Kip Thorne and one of his graduate students, in a 1988, but required many unknown features, and would certainly not be possible at the present time.

The principle involves time dilatation, which as noted above can be achieved in two ways. One end of the wormhole can either be accelerated to sub-light velocity (by some as yet unknown form of propulsion), or the mouth of the wormhole can be moved into a strong gravitational field.

Either way, time dilatation will cause that end of the wormhole to age less than the other, The wormhole then needs a spherical shell of exotic matter to keep it patent. Clocks at either end of the wormhole will always remain synchronized, but there will be a time differential between the time frames at the two ends, with the mouth of the wormhole effectively younger than the exit.

Hence, if the date in a prospective traveler's universe was 2014, for example, and at the opposite end of the wormhole 2020, once he traversed the wormhole, the time on arrival would still be that at which he entered the mouth of wormhole, i.e. 2014, though he would now be in a time frame for 2020 - in other words, he would effectively have traveled back in time by 6 years.

That nevertheless is a somewhat simplistic description, and there are many practical difficulties, some of which are unlikely to be resolved in the foreseeable future. Exotic matter for example can take many forms, none of them known at the present time, e.g. involving negative mass or dark matter particles; the energy requirements may also seem to violate certain 'energy conditions' of general relativity, though in fact, the amount of negative energy may be so small, that it might actually be possible, due to the Casimir effect in quantum physics.

'Adjusting' the mouths of the wormhole, to bring them sufficiently close to violate causality, on the other hand, might induce quantum and gravitational effects, which could effectively destroy the wormhole.

(3) Closed Timelike Curves (CTC)] . As note earlier, these are world lines which form a closed loop, such that any object which could move in this way, would eventually return to it's starting point. This is another example of something which is 'allowed' under the equations of general

relativity, though they have never actually been shown to exist, and may well not be possible at all.

Nevertheless, a variety of theoretical proposals[5] are based on these, such as the Tipler cylinder and traversable wormholes, though some physicists believe that these would not be compatible with a future theory of quantum gravity. Stephen Hawking even sees this as nature's way of preventing paradoxes and time travel abuses, by excluding time travel altogether on the macroscopic scale (chronology protection conjecture).

Although rather different, the spacetime fabric of the universe itself is curved, and perhaps it could function as a 'global' CTC, such that any random journey outwards 'into space', could in theory circumvent the entire universe and eventually return to where it started. Whether this would be within the lifetime of the universe or not, is another matter.

Tipler's cylinder was a hypothetical form of time travel, based on an infinitely long cylinder rotating rapidly about it's longitudinal axis. This produced frame-dragging effects, warping the adjacent spacetime in such a way as to produce closed timelike curves, which would allow an accelerating spacecraft to travel backwards through time. However, it was later shown to be impracticable, because the cylinder would either have to be infinitely long, or some form of negative energy made available.

A much simpler, 'scaled down' version of this might be a giant centrifuge, which could be accelerated up to arbitrarily high velocities, sufficient to produce time dilatation; unfortunately, to achieve the necessary velocity would be accompanied by 'unsurvivable' centrifugal forces. However, it might be possible to construct a 'working model', in which various 'levels' of organic matter (including microorganisms), could be studied in different time frames.

(4) Quantum Entanglement, between subatomic particles, allows widely separated events to be correlated in a way that suggests instantaneous communication between them. The no communication theorem[9], however, gives clear proof that entanglement cannot be used to transmit information faster than normal mean, and there is also no evidence that 'temporal sequences' are breeched with any of these effects[10].

The usual 'explanation' is that quantum states are linked (entangled) from the outset, and changes which occur are in some way predetermined, and therefore do not violate special relativity. Nevertheless, this does not explain the phenomena themselves, and although entanglement only exists on the quantum scale, certain aspects of the double-slit experiment, demonstrate very similar properties.

FTL travel enters into many time travel interpretations, but so far as physical processes are concerned, the ability to modulate, and therefore transmit information, is the all important criterion.

Nevertheless, there are still certain experimental situations which suggest 'reversed causality', for example, in the delayed choice quantum eraser experiment, where pairs of entangled photons are divided into two groups, and it can be shown that the way in which measurements are carried out, determines the information which the experimenter receives; in other circumstances, comparing measurements can suggest retroactive decisions, with respect to the eventual phenomena observed.

The more we study the quantum world, the more we find out, but the less we seem to know, and the more 'interpretation' comes to replaces 'exactitude'.

(5) Cosmic Strings are 1-dimensional defects in spacetime, which arise during symmetry braking between the Electroweak and the Hadron epochs, within 10-6 seconds of the Big Bang (Figure 4.1).

Such one-dimensional objects cannot be visualized in any normal way, but they have been likened to the cracks which appear when water freezes into ice[5].

Cosmic strings belong to a group of fundamental properties envisage by 'string' theory, and although no cosmic string has yet been found, theoretically they have a multitude of potential involvements in many different areas, including cosmology, inflation and brane cosmology.

Cosmic strings have no primary involvement in time travel per se, but could have great relevance to wormholes in this respect, since theoretically they can exist with negative mass. That might therefore provide the necessary form of exotic matter required to sustain the patency of a wormhole, and essential if time transport is ever to become practical through this means.

Additional to that, though stability of exotic matter will always be problematic, if a negative mass string was wrapped round a wormhole in the early universe, that might stabilize it almost indefinitely, and remove one of the main difficulties in trying to adapt a wormhole for practical use as a time machine.

Nevertheless, since any suitable wormhole will only be unidirectional, serious thought would need to be given to the return journey, before embarking in the first place.

13.7 Paradoxes of Time Travel:

There are a great many anomalies associated with moving through time, but undoubtedly the most 'general' of these, is why it may not be possible to travel into the past (which has at least existed) while almost certainly it is possible to travel into the future (which has not yet happened).

Hence, if time travel ever does becomes possible, that would strongly suggests a preexisting continuum of 'past', 'present' and 'future', within the spacetime fabric of reality, but nevertheless, continually accessed for that fleeting instant (which we experience as 'present') when 'past' transforms smoothly into 'future'. It is as though the 'space' and 'time' fractions of 'spacetime' only combined for that instant of 'present', which creates the physical reality of existence, of which we have awareness.

The following example might suggest why this is so. A time traveler, after completing a round trip journey 100 years into the future, leaves a written account of everything he found. After a further 100 years has passed by naturally, this record is then compared with what is now the real world, and found to be accurate. How could such a future be exactly as predicted 100 years earlier, unless it was either preordained or in some way, already existed?

There would be no equivalent anomaly for traveling back in time, however, because these events happened in our past, and we already know what they were. Nevertheless, the great majority of paradoxes arise in relation to traveling back in time[11], and the highly contentious issue of whether it would be possible to change past events (effectively before they happen) and thus alter the present.

The grandfather paradox succinctly illustrates this anomaly: a man travels back in time, and kills his own grandfather; hence, he himself could never have been born, to go back in time and carry out the act in the first place. Nevertheless, the principle is a general one, and would cover any action which would make it impossible to travel back in time in the first place.

The Hitler paradox is a similar example, in which WW2 might have been avoided, if someone went back in time and killed him as an infant;

by comparison with the grandfather paradox, however, it gives a graphic illustration of just how sweeping the consequences could be, from what in 'time travel' terms would be a relatively simple event to carry out.

No matter how unlikely, and not withstanding Hawking's proposals, until we have valid explanations for anomalies such as these, the possibility for misuses will always be there.

13.8 Solutions and Explanations:

A number of solutions have been suggested for most individual paradoxes - in fact frequently more than one for any given paradox -ranging from parallel universes (multiverse), to laws of nature which would forbid time travel altogether, such as Stephen Hawking's chronology protection conjecture, and it is that element of uncertainty which makes the question of paradoxes both fascinating and frustrating.

Ideally, it would be nice if we could generalize, with one explanation for traveling into the future, and another for traveling into the past. Unfortunately, whether something is possible or not, usually depends on individual proposals, and since there is always likely to be more than one way of achieving any given objective, anything which nature might forbid is still only relevant if it is actually being used in practice.

Temporal paradoxes themselves have been used to argue that time travel must be impossible, because it is capable of resulting in a paradox. However, Kip Thorne, has argued that none of the supposed paradoxes formulated in time travel, can actually be formulated at a precise physical level, i.e. whatever the paradox may seem to be, it is always possible to show that there are many consistent solutions[11].

Nevertheless, not everyone would go along with that, preferring instead to accept a range of options, out of which perhaps one will explain one paradox, but not another, while a different explanation may explain a different paradox, in such a way that there "must always be an answer" - provided there are enough options to choose from, and the following have all been propped:

Muliverse (3.6). This postulates an infinite number of almost identical universes, and that by going back in time, a traveler would enter one of these, and then proceeds to kill his grandfather ; on returning, he would re-enter his own universe and carry on with his life, because the grandfather he killed was not his own biological predecessor.

This is a strange 'explanation', because the act of 'changing' universe, is at least as puzzling and requiring an explanation, as the paradox itself.

Branching Universes. This has a very similar basis, and proposes that traveling back would cause time to branch, with different universes having different futures, and again, it would not be his 'real' grandfather that he killed.

Timeline corruption. This is rather more general, and would not explain the grandfather paradox. Rather, it rests on 'butterfly effects', in which a traveler could cause many different small changes, and these would then spread out into the future, and produce their effects there. They might also interact or cancel out, and in the long term, the 'future' would be entirely different from what it might otherwise have been.

Timeline Erasure. The time traveler can change the past, erasing one timeline and existing in another, and the dead grandfather would again not be in the travelers future timeline.

Merging timelines. With these, actions committed in one timeline overlap and merge with those of another, for example, if the traveler met up with a 'double' from another timeline, they would merge, and the traveler would then belong to that timeline. Any two 'incompatible' events (e.g. the grandfather) would always merge into whichever alternative timeline would not produce a paradox.

Choice timeline hypothesis - the act of deciding to travel back in time changes history, thereby rendering his actions in that regard pre-destined.

Self-healing hypothesis - If a person decides to go back in time, to change the past in order to alter the present, there will always be another set of events that cause the present to happen, and so it would be pointless going back to try to change it.

The Novikov self-consistency principle argues that whatever the traveler does in the past was already part of history, and it is impossible to change history. Any attempt to kill his grandfather was therefore bound to fail.

13.9 Life Extension:

This is not Time Travel in any accepted sense, but it is a theoretical possibility today, and there is ample precedence for such things becoming the reality of tomorrow - in which case it is a perfectly feasible solution to extending journey times, beyond what would be possible within a normal human lifetime.

It would no doubt involve complex preparations of some sort; these might even be semi-long term in themselves, such as a routine of drugs, diet and life-style, though to be practical, there would have to be an 'achievable endpoint', at which a person was finally prepared and ready to undergo the procedure itself.

It is not possible at present to say what that might involve, though probably some form of 'extended sleep', anaesthesia or, artificial hibernation, which we know to be feasible in animals for long periods of time, and therefore at least physiologically possible.

There are two ways that this might be accomplished in practice - 'Artificially' or 'by Design'. The former would involve placing a person in a state of 'unawareness' - not necessarily unconscious - with metabolism and respiration suppressed to a minimum, but oxygen would still have to be available in some form, for the definitive cause of all deaths is cerebral hypoxia[12] (irrespective of any associated factors).

The alterative, and preferable option, would be to find out why biological entities age in the first place - a progressive decline throughout life, and eventually death from old age. However, there are some creatures which never die of old age e.g. hydra, planarian flatworms, and certain jellyfish[13], and if we could establish how that comes about, there could be any number of potential benefits, quite aside from 'life extension' and extended space travel..

Such creatures are biologically simple, however, compared to human beings, and almost certainly it is complexity - cellular diversity of structure and function - which precludes the practicality of 'total survival', beyond some arbitrary point, which cannot be extended indefinitely.

This 'life profile' in turn must have been programmed into genes at the outset, and ideally we need to find out which ones are responsible for biological decline and eventual death, and see if they can be removed.

However, if these also subserve other purposes, then the alternative option might be partial removal, or selective manipulation, which would amount to manipulating DNA itself.

These are not the only relevant factors, however, and not everyone might want or agree to the development of 'life extension'. In a recent U.S. study, for example, 56% said they would reject life extension, and only 4% would consider 120 years to be ideal[14]; the oldest documented human life is 122 years[15].

It took human life almost 800 million years (6.4) to evolve to it's present complexity, however, and clearly in doing so, certain primitive advantages had to be sacrificed along the way; the question now, is whether any of these can be reinstated?

Genetic engineering is still in it's infancy, but manipulating DNA will almost certainly become possible at some future time.

Reference (Chapter 13):

1. E. P. Mitchell, The Clock That Went Backward (New York Sun, September 18,1881)
2. M. A. Bodin, Primer of Relativity (Trafford publishing, 2006)
3. A. Einstein, Relativity: the Special and General Theory (London: Methuen & Co Ltd, 1962)
4. Wikipedia article (! December 2013), Cherenkov radiation.
5. Wikipedia article (1 December 2013), Time travel - Time travel to past in physics.
6. J. Hawley and K Holcomb, Foundations of Modern Cosmology (new York: Oxford University Press, 1998
7. Google: Do Tachyons Exist?
8. J. Gribben, Q is for Quantum (Weidenfeld & Nicolson,
9. Google: What is no-communication theorem?
10. L. Bengtsson, K, Zyczkowski, Geometry of Quantum States. An Introduction to Quantum Entanglement (Cambridge University Press.2006)
11. Wikipedia article (1 December 2013) Paradoxes of Time Travel.
12. Wikipedia article (26 July 2014), Senescence
13. P. A. Newmark, A. A. Sánchez (2002). Not your father's planarian: a classic model enters the era of functional.
14. http://www.pewforum.org/2013/08/06/living-to-120-and- beyond-americans-views-on-aging-medical-advances-and-   radical-life-extension/
15. C. R. Whitney, Jeanne Calment, World's Elder, Dies at 122 (The New York Times, 5 August 1997)

Chapter 14

ALIEN LIFE

(Sp eculative biochemistry)

14.1 Nomenclature:

Biochemistry is the chemistry of life - from unicellular organisms to the complex multicellular structure of human beings, and all depend on three features:

Carbon - for structure and metabolism (organic chemistry).
Water - as a solvent.
Nucleic acids - for data and information handling.

Molecules can be described both in terms of their formula and their structure, while those with the same formula but different structures are known as isomers - i.e. contain the same number of atoms of each element, but with different spatial arrangements. Isomers do not necessarily share similar properties, unless they also belong to the same functional groups.

Enantiomeres are one of a pair of steriometric isomers, which are mirror images - i.e. opposite orientation, commonly designated as left (L) and right (D) handed - and therefore cannot be superimposed. Racemic mixtures have equal amounts of both.

Easily confused are chiral molecules, which have a non-superposable mirror image. The presence of an asymmetric carbon atom is often the feature that causes chirality; achiral objects, are symmetrical, and identical to their mirror image.

Ribosomes are large complex molecules, characteristic of all living cells, as the primary site of protein synthesis, and which link amino acids together in the order specified by messenger RNA.

14.2 Introduction:

Alternative Biochemistry includes all forms of biochemistry that are scientifically viable, but so far as we know, do not exist in nature at the present time, though theoretically might do so elsewhere in the universe.

That would be under different environmental conditions, and could, for example, be on the surface of some exoplanet or satellite, within it's atmosphere, or even in the environment of space itself.

Such alternative life forms have long been the prerogative of science fiction, whether humanoid or alien in characteristics, though most of those commonly seen (e.g. Star Trek) usually seem quite happy in an oxygen environment, and could very well be organic - bearing in mind that the chemical composition of the universe as a whole is heavily weighted in favor of carbon based life.

The possible existence of other life forms can never be ruled out, but there are a number of background factors, concerning the existence of life in general, which might have a bearing on that.

These relate to the assumption that life may have a purpose (intelligent design) and is not just an arbitrary eventuality, which is a logical conclusion from 'fine tuning' (7.3). In that respect, having designed a universe to ensure, for whatever reason, the development of carbon life, and it's eventual evolution to intelligence, what would be the purpose of introducing yet another form of life?

It is difficult to think of any advantage, and we would strongly disagree with Drake's estimate that 100% of all life forms which evolve elsewhere, would progress to become intelligent (11.7).

Since organic life already exist (on earth), and shows every prospect of escalating to achieve almost any foreseeable purpose, given sufficient time - and the geometry of the universe provides for that - the only logical corollary would be the widespread evolution of organic life elsewhere, rather than some other form.

Such an argument, nonetheless, simply brings us back to Fermi, and the reasons why we have not been able to confirm something which logic tells surely must exist.

However, we can never exclude the possibility of life in other regions of the universe, where carbon may not be common, for our assumptions of uniformity, which seem both necessary and justified in that 4% of the universe that we can see and observe, may not be typical of the universe as a whole.

Speculation about alien life dates back to antiquity, but practical considerations as to it's nature and physical form, are much more recent. Carl Sagan in particular did much to popularize extraterrestrial life, both in science fiction and in the reality of his professional life, where he coined the phrase "carbon chauvinism" for the questionable assumption that what applies to life on earth will necessary apply elsewhere in the universe.

This possibility has been embraced by science fiction for years, and it was the need for aliens to be different that provided the incentive to come up with an alternative chemistry, and silicon seemed a likely choice.

Given the importance of that role, however, and silicon has assumed a somewhat exaggerated status over the years, while the reality is, even if it is the most likely substitute for carbon, it is still barely credible as a realistic alternative to organic life, and from what we do know of it's chemical potentials, advanced intelligent life, would no more be within the capabilities of silicon than it would for any other element.

Nevertheless, there are many proposals for non-carbon biochemistries which do have to be considered before they can be ruled out, and the simplest of these would not even be chemical, in the strictest sense, but a property of symmetry rather than composition.

14.3 Chirality.

It is a matter of debate, nonetheless, whether steriochemistry[1] can really be regarded as alien or not, and molecules that may be predominantly of one enantiomere in one particular organism, can still be found in the opposite enantiomere, in different organisms, particularly among the unicellular Arachaea.

In higher forms of life, amino acids are almost universally L-enantiomeres, while sugars, by contrast, are nearly always in the D form. However, molecules of opposite chirality still have identical chemical properties to their mirror equivalents, so that theoretical life comprising D-enantiomeres of amino acids and L-entaiomeres of sugar might exist; and while these would be mutually compatible, they would not be compatible with their 'real life' counterparts.

Nevertheless, all of these are still organic, and it is non-carbon based life, and whether that might exist elsewhere in the universe, that is really of interest.

14.4 Properties of Carbon:

The element Carbon is the basis of life, yet as we saw earlier (7.7), it owes it's existence to a physical property (resonance) which is so improbable, that organic life remains one of the great unexplained mysteries of nature.

Nevertheless, carbon is now the 4th most abundant element in the universe, the 15th commonest element on earth, and the second most abundant element in the human body[2].

It is carbon's unusual polymer-forming abilities, however, with 4 covalent bonds, each with a carbon-carbon binding capacity twice that of the equivalent silicon bonds, that make it ideally suited as the chemical basis for living entities.

20% of the carbon in the universe is also now thought to be associated with polycyclic aromatic hydrocarbons (PAHs), which may have been abundant in the primordial soup, while recent findings suggest that PHAs in interstellar space can transform into more complex organic compounds, used to form amino acids and nucleotides (8.6,8), which in turn are the raw material of proteins and DNA.

14.5 Silicon as a basis for Life:

Silicon is the most likely element that could replace carbon in the biochemistry off life[1], and has many basic properties similar to those of carbon. It is in the same group of the periodic table, and like carbon can also form the large molecules necessary to carry biological information.

The earth (and other terrestrial planets) are silicon-rich, and carbon-poor, with a silicon to carbon ratio in the earth's crust of almost 900:1. Nevertheless, surprisingly perhaps, silicon is rarely found in biological material.

The reverse is true in interstellar space, however, where carbon molecules greatly outnumber those of silicon, and overall, the cosmic abundance of carbon to silicon is roughly 10:1[3].

These statistics, in themselves, indicate why silicon is less suitable than carbon for biochemical purposes, but not only is it less common in space, it is also much less versatile in forming other compounds, those which do from tend to be unstable, and overall there is a much greater variety of carbon than silicon compounds.

Hence, silicon has some significant disadvantages as a potential basis for life:

- It interacts with few other atoms.
- Silicon atoms are too large to readily form the necessary double bonds characteristic of organic chemistry.
- Silicon is relatively uncommon in interstellar space, where the initiation of life takes place.

Nevertheless, it is perhaps unwise to draw too many parallels with carbon life, which is specifically suited to the environments in which it exists. These certainly cover a very wide range of conditions, and on that basis, any form of extraterrestrial life might well be similar to our own - under comparable environmental conditions.

In the widest sense, however, what we are really interested in is whether any form of life could evolve and adapted under some of the totally different conditions which prevail elsewhere in the universe.

That said, however, and we are still left with the question of 'fine tuning' (7.3), which would seem to suggest an inevitability for organic life to exist, and also raises the question of just how 'statistically random' the workings of the universe really are.

From these sorts of considerations, it is impossible to rule anything out, but on balance overall, silicon based life seems unlikely.

Nevertheless, silicon compounds may still be biologically useful under different conditions of temperatures and pressures than occur on the earth's surface, and it has been suggested, for example, that the first complex organic molecules to exist on earth may have formed out of clay minerals, based on pre-existing inorganic silicon crystals - 8.5(5).

It was once even suggested that the silicon based microprocessors of early computers, might be classified as a primitive form of alternative life, and though such an idea may seem ridiculous in this day and age, things have now almost progressed full circle, to the foreseeable possibility of artificial life (6.7) assuming a dominant role in the long term future, with silicon perhaps at the basis of an eventual technological singularity (15.8).

Should such a situation ever come about, however, the eventual reality in fact would be a hybrid, as much dependent on carbon as silicon, while organic life per se would face a very different future, once it became redundant.

14.6 Other Possible Chemical-based Life[4]:

Metals: A number of metals, such as titanium, aluminium, iron and manganese, which are all commoner on earth than carbon, can combine with oxygen, to form very complex structures, analogous to those of organic compounds..

Metal oxides also have a number of properties similar to carbon, and theoretically might provide an alternative to carbon- based life in situations where that might be impractical, for example high temperature environments.

Wolfram (Tungsten),which occurs naturally in compounds, is another possibility, because of it's robustness, high melting point, and a remarkably high density, almost 20 times that of water. A group of biologists at Glasgow university have created life-like cells based on tungsten polyoxometalates[5]

Sulfur is another element capable of forming long-chain molecules, though these do not branch out in the way that carbon chains do. Biologically, it has long been known as an electron acceptor, and sulfur-reducing bacteria use sulfur in place of oxygen.

Arsenic is involved in the biochemistry of a number of organisms, while others can utilize arsenite to provide energy, and it has been suggested that earliest forms of life could have used arsenic in place of phosphorus in their DNA structure[6].

14.7 Alternative Solvents:

All known forms of terrestrial life use water as a solvent, and indeed the suitability of an environment to sustain liquid water is one of the essential criteria for habitable zones (5.10,7.5), and therefore could be an important consideration where possible extraterrestrial life is concerned.

Properties which make water important for life, include the large temperature range over which it remain liquid, ability to dissolve a wide range of substances, being less dense as a solid (ice) than liquid (10.6), and it's ability to act either as an acid or a base - crucial to many biochemical reactions.

Not all properties are advantageous, however, for example it's high reflectivity as ice, which can significantly reduce the light and heat received from the sun.

The possibility of extraterrestrial life based on solvents other than water[1], is topical with astrobiologists at the moment, and alternatives considered have included, ammonia, hydrocarbons, methanomide and even sulfuric acid and liquid nitrogen.

Of these, ammonia, is probably the most realistic, and was first suggested by J. B. S. Haldane in 1954[7], because of it's many similarities to water. It is almost as abundant in the universe as water, can act as either an acid or a base, dissolves organic molecules just as well, and many common water-related compounds also have ammonia-related analogs.

As a basis for life, however, there are several difficulties. Primarily, it is inflammable in oxygen, and so would be incompatible with any form of aerobic life. It could have been present in the early prebiotic environment, however, where reducing conditions existed (8.3), but wether it could hold prebiotic molecules together long enough for self-replication, is questionable.

Ammonia based life would also have to be in a much colder environment than terrestrial life, where ammonia would vaporizes at a much lower temperature than water, and both metabolism and evolution would also be slowed down.

In general, to be a successful substitute for water, any solvent must remain liquid over the wide range of temperatures that might be encountered, e.g. on the surface of exoplanets, though because boiling points vary with pressure, the level of that (rather than temperature), below which a solvent would still remain liquid, might be a better criterion.

Nevertheless, pressure is a factor which would vary so widely, that rather than a single alternative for extraterrestrial life, a number of possibilities might have to be considered, in relation to different environmental circumstances.

Hydrocarbons are a group of organic compounds which might also provide an alternative solvent to water. Methane, for example, has a similar cosmic abundance to ammonia, and lakes containing methane (among other chemicals) have been found on Titan, by the Cassini space probe.

Chris McKay, an astrobiologist, has suggested that if life did exist on Titan, it might also use more complex hydrocarbons as an energy source[8], and the higher abundance of molecular hydrogen in the upper layers of Titan's atmosphere compared with lower, identified in 2010, would suggests a redistribution of hydrogen molecules compatible with that sort of methanogenic life-form.

Nevertheless, other explanations for these findings are considered more likely, such as non- biological catalyst effects - though that in itself would be a significant discovery.

Hydrogen fluoride has a number of properties comparable to water and ammonia, but unlike these, it is rare in the universe, while Hydrogen sulfide is the closest chemical analog to water, and though not common either, it is quite plentiful in liquid form below the surface of Jupiter's moon Io.

Liquid Sulfuric acid is abundant as droplets in the thick atmosphere of Venus, and with two carbon atoms joined by a double bond, could provide an analog to water-based biochemistry.

Nevertheless, given all the above pros and cons, and clearly there is nothing remotely approaching an alternative to water as universal solvent - which must greatly reduce the likelihood of any other forms of biochemical life existing elsewhere, because the number of individual circumstances required would make it impractical for nature to evolve an equivalent diversity of life forms, especially in an environment already preordained to support organic life.

14.8 Alternative Environments:

The earth's atmosphere underwent a major transition in the later stages of the prebiotic era, about 2.4 billion years ago[9], when photosynthetic Cyanobacteria began producing oxygen, which wiped out the preexisting population of anaerobic organisms (6.5- oxygen catastrophe), and permanently changed the atmosphere from reducing to oxygen rich, which has persisted ever since.

These changes impacted both on biochemistry and morphology of existing life[10], and even today, many plants and animals can undergo major biochemical changes during their life cycle, in response to changing conditions - a versatility of existing life, but which might well be inherent in more exotic forms of life too, in an attempt to match alternative biochemistries with the diversity of atmospheric and other environmental conditions prevailing elsewhere.

14.9 Alien Visitations

A number of people genuinely believe that aliens are here among us at the present time, disguised in human form, and therefore indistinguishable from ourselves.

There are two other ways, however, in which alien visitations to earth may have taken place, which were not so anonymous, and in fact there is abundant indirect and anecdotal evidence for both, though how reliable that evidence may be, is another matter: Unidentified Areal Phenomena (Flying Saucers, or UFOs); and Ancient Astronauts, visiting earth in pre-historic times.

(1) UFOs are not new, and lights and strange objects in the sky have been regularly reported throughout recorded history; their prevalence today, however, probably owes more to emotive terminology, than any change in frequency.

'Flying saucers' were first reported in 1947, and very quickly acquired a reputation for 'alien interference', which governments could hardly ignore, in spite of their denials, and the military certainly took seriously.

There have been many subsequent accounts of landing and abduction, over the years, but no formal attempts to make contact, as would be expected from any intelligent beings who had traveled so far, and apart from the questionable Roswell incident, which has now largely been discounted, we know nothing about them at all.

It is this unwillingness to reveal themselves, other than in a very surreptitious way, which places the greatest doubt over the reality of their existence, and there are so many other inconsistent features, that as a race of beings, we must assume they do not exist.

Likewise, the sightings and phenomena of the 'saucers' themselves can be accounted for in so many ways, that their reality as alien spacecraft can also be safely dismissed.

With hindsight, and free access to official records, most of the glamor has long since been laid to rest, though if, as we are led to believe, there are still a small number of incidents which cannot yet be explained, these too should at least be put in the public domain - otherwise one is left with the impression that something which does matter, is being withheld.

Scientists, officially at least, have always been skeptical about contact with alien life, but the reality of it's existence elsewhere in the universe is now widely accepted. Plans already exist to upgrade SETI, for example, though whether these will ever be funded, is the great uncertainty with applies to all projects of this sort.

Nevertheless, the very recent findings of 'Kepler' as to just how common planets like our own earth actually are, can only be helpful in this respect.

(2) Ancient Astronauts: The myths and legends of antiquity are filled with records and descriptions of strange and unfamiliar objects, relics and artifacts of every sort, from all corners of the earth. These cover many situation and circumstances, but implications of space travel are common to many; pictures and wall-paintings, for example, showing humanoid beings in space suits, their limbs disposed as though in free-fall, or some other weightless environment, such as EVAs.

Many objects are unfamiliar, while others are clearly anachronistic, such as helicopters and submarines, or tools and machines, which had not yet been invented, and therefore must have been acquired from elsewhere.

Pseudo-religious objects and figures abound, such as winged humanoids, or angels, together with written and symbolic texts, and depictions reminiscent of miracles or biblical events, while others resemble Gods or deities of some kind,.

Some objects are not easily recognized, but can be interpreted as familiar, though these are not always consistent, and this is one difficulty with such artifacts, which can be described and 'explained' in different ways by different people.

It is difficult to generalize this sort of material, but 'religion' and 'spaceflight' are common themes, as are anachronisms, and a number of features resemble those which exist today, such as pyramids, obelisks or in relation to tombs and coffins.

The ancient astronaut hypothesis attributes all of these to extraterrestrial beings who visited earth and made contact with early man in the antiquity of prehistory, and thereby influenced basic features of human development, such as culture, learning, technology, and overall that many abilities and accomplishments which might otherwise suggest divine intervention in some form or other, were in fact obtained from those of advanced extraterrestrials.

A number of authors have specialized in these areas, and published there own theories and overall interpretations.

Erich von Daniken is perhaps the best known of these, and has attributed a wide range of artifacts to knowledge or skills imparted by ancient astronauts, including the Great Pyramid of Giza, Stonehenge, the monolithic human figures on Easter Island, and a number of Mesopotamian artifacts, for example suggesting galvanic cells. His book, Chariots of the Gods, was an international best seller[11].

Most religions, he believes, show evidence of 'alien input', and that supernatural events, for example, might simply reflect more advanced alien technology, while common features of geographically widely separated cultures, are suggestive of common origin.

The exact chronology of alien astronaut theories is not discussed by any of it's proponents, so we really have no idea of just how old these events actually are. It has been suggested, for example, that Darwin's theory of natural selection is inaccurate, because he may have underestimated just how much time the early stages of human evolution would actually take.

Alfred Russel Wallace has suggested that some of the gaps in evolution were due to the intervention of a 'creative spirit' rather than natural selection, but without a shred of evidence to back it up, it must be one of the least likely criticisms of Darwin's work, among others which at least have a semblance of credibility.

Carl Sagan believes that the possibility of extraterrestrial contact during earl human history, though purely speculative, is nevertheless worth taking seriously; he believes that sub-light velocity interstellar travel would have been inevitable, just as soon as it was developed.

Sagan also believes that ancient records are a potentially reliable way of establishing visitations, though goes on to point out that some of these might have been written later, after the event, or could be second hand, in whish case 'embellishment' could well be a feature.

It is also worth noting that after the amount of time already devoted by von Daniken, among others, to tracking down and analyzing information of this sort, there cannot be many worthwhile 'sample' which have not already been considered.

The Medieval Graffiti Project[12], by contrast, was started by medieval archeologist Matt Champion, in 2010, to collect graffiti from church premises, and now has over 28,000 images from all over the country. The variety is large, both figures, objects and text, while a small number are not easy to identify.

In Mr Champion's words "There are a variety of different theories and care is needed when interpreting the drawings"12].

Though very different from the imagery of von Daniken and others, such a project has unknown potentials, and is well worth watching as it spread across Europe.

14.10 Shadow Biosphere:

This was a concept first introduced in 2005, which proposed a hypothetical terrestrial biosphere of microorganisms, using radically different biochemical and molecular processes from those of carbon-based life[13].

Their presence would be so obscure as to go entirely unnoticed, and their very existence unsuspected, given the universality of organic biochemistry with respect to all other known forms of life.

Nevertheless, it was suggested that if organisms based on RNA, which may have existed early in the prebiotic era, could still be alive today, they would have gone unnoticed because they do not possess ribosomes (14.1), which are conventionally used to identify living organisms. Clearly they could only exist under very unusual conditions, such as in tiny mineral pores, less than one micrometer in size, or extreme temperature environments[14].

Other proposals have included amino acids and nucleotides with chirality opposite of ours, arsenic replacing phosphorus[15], and desert varnish[16] - a dark brown coating found on the surface of rocks in arid desert conditions - whose status as living or nonliving has been debated since the time of Darwin.

None of these have been confirmed however. Attempts to do so can be criticized on a number of grounds, and the concept of such a biosphere is not widely accepted.

14.11 Artificial life:

We have looked at some general aspects of this in chapter 6.7, but it's inclusion here reflect the fact that it will undoubtedly become a reality at some future date, once human beings have developed the technology for interstellar travel, and themselves begun to colonize the galaxy.

Whether it exists already in the universe at large, depends on whether extraterrestrial beings exist, with sufficiently advanced technology - which is something we have yet to establish - but whether it does or not, it is inevitable at some stage (15.3).

Once that has been accomplished, the future is reassured. It will only be a matter of time before artificial life becomes established as the dominant form of extraterrestrial life, and we discuss that in the next chapter.

References (Chapter 14):

1. Wikipedia article (29 June 2014), Hypothetical types of biochemistry.
2. Wikipedia article (24 June 2014), Carbon..
3. Wikipedia article (15 June 2014), Abundance of elements in Earth's Crust.
4. In reference 1: Other Exotic Element-based Biochemistry.
5. Life-Like cells are made of metal (New Scientist, September 14, 2011).
6. M. Reilly, Early life could have relied on 'arsenic DNA' (New Scientist,198, 26 April 2008).
7. J. B. S. Haldane, The Origins of Life (New Biology 16: 12–27. 1954).
8. C. P. McKay, H. D. Smith, Possibilities for methanogenic life in liquid methane on the surface of Titan. (Icarus 178 (1).
9. B. W. Jones, The Search for Life Continued: Planets Around Other Stars (Chichester, UK: Praxis Publishing, 2008).
10. P. G. Falkowski, and others, The Rise of Oxygen over the Past 205 Million Years and the Evolution of Large. Placental Mammals (Science 309, 30 September 2005).
11. E. von Daniken, Chariots of the Gods (Econ-Verlag, 1968).
12. Medieval Graffiti Project (Internet: BBC News, 20 July 2014).
13. Wikipedia article (5 July 2014), Shadow Biosphere.
14. S. A. Benner, A. Ricardo and M. A. Carrigan, Is there a common chemical model for life in the universe? (Curr. Opin. Chem. Biol., 8, 2004)
15. P. C. W Davies, S.A.Benner, C. E Cleland, C.H. Lineweaver, C. P. McKay and F. Wolfe-Simon, Signatures of a shadow Biosphere (2009) (Astrobiology.9(2), 2009).
16. Wikipedia article (14 April 2014), Desert Varnish.

Chapter 15

EXTRATERRESTRIAL LIFE

"Where are they"?
- Enrico Fermi

There is no clear record of when the term 'extraterrestrial ' was first used, but the concept of life existing elsewhere in the universe is not new. By contrast, only very recently have we acquired the technology to identify and study other planets, in sufficient detail to make decision regarding their suitability, or otherwise, as habitats for life (12.10).

15.1 Introduction:
The reality of extraterrestrial life has always assumed the existence of suitable conditions elsewhere, and while these have primarily been visualized in terms of conditions on earth, the possibility of other life forms based on alternative biochemistries,, (chapter 14) and perhaps existing under very different conditions can never be ruled out[1].

We can say nothing about the suitability or these, however, without some knowledge of what that biochemistry might be, but from the variety of life which does exist, and the extreme conditions under which some forms can survive (8.6), it would be difficult to exclude almost any conditions absolutely.

Exoplanets (12.10) have always been presumed to exist, but they are difficult to identify, and until recently, impossible to study individually because of proximity to their parent star.

All that has changed, however, since the launch of the Kepler space observatory in 2009 (12.9). This is now identifying new planets on an almost daily basis, and can also analyze physical features, such as temperature, pressure and atmosphere (12.9). Biomarkers, for example oxygen - which depletes quickly unless life is present to constantly renew it[2] - are particularly importance, and among key criteria for the suitability of life.

Astrobiology (exo-biology) is a comparatively recent discipline of cosmology, which deals with all aspects of life in relation to the space environment, including the adaptability of human physiology, exposure of primitive life forms to space environments, and the search for microbial life on other planets, e.g. the surface of Mars (12.7).

Other objects currently being studied include several of the satellites of gas giants, larger asteroids, and comets. As technology advances, ever smaller bodies can be explored, with 'Sample Return' missions the eventual objective, providing a wide range of specimen to analyze for possible life

Other than SETI, however, all practical research into extraterrestrial life is at this level of nature, though there is little glamour attached, and few people appreciate the implications of this sort of work.

The Allan Hills meteorite, however (9.4), did seem like a genuine 'first' and received world wide attention, but there is nothing like uncertainty to dampen enthusiasm, and interest in this has long since declined.

Extraterrestrial life owes it's prominence largely to science fiction - previously books and films, but now mainly TV series, such as Star Trek - where it has become almost synonymous with humanoid aliens; benign integration with humans (e.g. crew) also reflects an emphasis not present in the past, and which belies the more traditional image, based on the Frankenstein era.

It was flying saucers (14.9), however, which did more than anything to bring alien life into public prominence, where it has remained intermittently ever since; concerns over alien surveillance, have now largely subsided, while 'abductions' remain anecdotal and have never been substantiated.

Eric von Daniken (14.9), in the late 60s, and his theory of ancient astronaut visitations, was the next peak in public interest, combining novelty value with existing artifacts, and still has a substantial following today.

The image of alien life has fluctuated over the years, but man's progressive expansion into space has brought the question of eventual contact into the bounds of possibility, and overall, a general acceptance that it will probably happen at some stage - including many professionals, who hitherto have tended to be skeptical.

Nevertheless, SETI by and large has made little impact on public interest - after 50 unsuccessful years, hardly surprising - and certainly done little to foster the image of intelligent aliens, more technologically advanced than ourselves, while Fermi's 'Silent Universe' remains as much an enigma as ever (15.6).

15.2 Background:

The 'many worlds' concept is a philosophy common to most ancient cultures, and in contrast to the reserved skepticism which prevails today, the existence of other beings was often taken for granted. Knowledge of the universe at large was non excitant, however, while the sun, moon and planets were all visualized in terrestrial terms and thought of as inhabited.

While all cultures had their own brand of 'alien beings', there was an underlying normality, nonetheless, and in many cases, such beings were regarded more as we would regard foreigners from another country - 'similar but different' - and indeed it was only natural if celestial bodies were assumed as similar to earth, that they would likewise be inhabited by essentially similar beings.

The concept of life elsewhere, in it's present context, relates to the universe as we now know it to exist, and therefore precludes all of the above, though the term 'extraterrestrial' is not really appropriate for the microbial life forms that are our main focus of attention in the solar system today.

There were many variations between these two extremes however, and inevitably religion, with it's core beliefs about God (who "created man"), and existence in general, has always had an influence on all questions of life. Nevertheless, it has never expressed specific views about alien life; and though the bishop of Paris in 1277 reputedly conceded that God might have created more than just one world, he was scrupulously reticent on all questions of inhabitants.

The Ptolemaic, earth-centered solar system (and universe) has always been an important tenet of church belief, and the introduction of the Copernican system was a heavy blow. Almost over night earth lost it role of central importance, and it's status was reduced to that of 'one among many'.

Coupled with the invention of the telescope a few decades later, man's whole concept of the universe changed irrevocably - so large and with so many other 'bodies' - that perhaps life was not that unique after all.

It was a logical shift of philosophy, and with it the concept of extraterrestrial life took on a pro-rata perspective - and sometimes rather literally, when one astronomer even suggested that if Jupiter was inhabited, such beings would be larger than humans, in proportion to the difference in size between the two planets!

Ideas about Extraterrestrial life soon became both common and acceptable. William Herschel, for example believed that the solar system, among others, was populated with aliens[3], and by the 19th century, life elsewhere in the universe was assumed to be common (cosmic pluralism), even including the sun[4].

Most of these earlier views seem to accept that alien life would be essentially similar to humans, while many extraterrestrials today are anything but. It is not clear when this distinction first arose, but Nicolas Flammarion, the French astronomer is often credited with highlighting the difference[5] - that aliens were a species of their own, and not simply a variant of human life.

The turn of the 19th century was the last serious reference to intelligent aliens in the solar system, when Percival Lowell proposed his theory of life on Mars (12.4).

The 20th century marks man's entry into space, initially overlapped by flying saucers, and later von Daniken, but by the turn of the 21st century, these were history, while planetary surface explorations were well under way.

Humanoid aliens, however, if any do exist, are either not listening to their radios, or just disinterested, and the reality of these, together with Fermi's paradox, are both in abeyance for the foreseeable future (15.6).

15.3 Life in the Universe:

What sort of life might exist elsewhere? We talk glibly about life as though it's existence was common knowledge, while in fact the only certainty about life, is the reality of our own existence.

Every time we speak of life, no matter the context, we are making the unspoken assumption that it is similar to our own, and even when discussing extraterrestrial life, while admitting that it could well be very different, we can still only theorize in anthropomorphic terms, and it is prejudices such as these which are going to make finding life (if it does exists) that much more difficult.

Hence, before we start looking elsewhere, we should do two things: accept one assumption - that nothing is impossible, until we have proved categorically that it is; and look carefully at the existence of Extremophiles (8.6), whose presence was overlooked for so long, because of preconceptions.

They may be tiny by comparison, but they live and tolerate conditions that would be instantly lethal to any human being, and they will no doubt survive long after man has either moved on, or progressed to shared existence with his mechanical 'offspring'

Nevertheless there must be physical environments in which even extremophiles could not survive - but with hind sight, probably not many.

Inside the fusion core of a star might be one, and certainly within the event horizon of a black hole, but otherwise there are probably few circumstance where they could not exist.

To be more specific, organisms of that sort have been shown to survive boiling water, frozen in a block of ice[6], within a nuclear reactors, liquid helium, acids and in interplanetary space. That is not the same as saying that life could arise de novo in any of these situations, but we have been wrong before, and it would be unwise to speculate again.

It is intelligent life, however, which is of greatest interest, and while contact might be by SETI, given it's track record, and that seems improbable; much more likely, contact will be initiated by aliens, rather than man, and at a time and a place of their choosing rather than ours.

Intelligent Aliens are undoubtedly the biggest uncertainty of all, but having reviewed alternative biochemistries in the previous chapter, we consider organic life as the only likely form of chemical life elsewhere in the universe.

That does no preclude artificial life, or as a practical reality, artificial intelligence in control of mechanical technology.

Intelligence, as we have seen, is an evolving property - innate within human beings. However, it is also the property which will maximize the efficiency of man's use of technology, and it is technology which man hoped would take much of the burden of existence off his hands. But "the best laid plans of mice and men.......", and this time man got it wrong.

He started off alright, and was able to produce a program, with built-in safe guards, to ensure that 'clever machines' would not become 'too clever too quickly', and take over control.

But he overlooked one thing - Moore's law, that technology evolves about 3,000 faster than human intelligence (6.8). Within a very short period of cosmic time (about 100 years), artificial intelligence had outstripped human intelligence, man soon became redundant, his powers of veto overruled, and for the first time in history, machines were now in total control (Technological Singularity)[7]; thereafter, they remained in control for the lifetime of the universe, while man's future was very different from what he had planned and expected (15.8).

15.4 Searching for Life:

From a practical stand point, we can only search 'probable environments with current technology', and we are looking to confirm 'past or present existence', with, again, some uncertainty regarding criteria which may never be resolved. The Allan Hills supposed micro fossil is an example, where current technology is not sufficient, while one scientist has stated that "it can be virtually impossible ever to distinguish micro fossils unambiguously from artifacts".

Nevertheless, micro fossils will always be one of the commonest specimen types available to us, and hence it may prove very difficult ever to get categorical confirmation of extraterrestrial life based on samples of this sort.

The actual sites for such specimens will be 'surfaces' (of any suitable body) including samples obtained through drill holes, perhaps to a depth of a few feet; 'gaseous' e.g. samples obtained within planet atmospheres, and 'liquids' of any sort which are physically accessible, but water/ice in particular. Some of the Jovian satellites may have subterranean lakes of water, and these are now priority sites for soft landings.

Man will never give up his convictions of intelligent aliens, however, and therefore 'looking' for these will continue, though in practice that means 'listening'. Nevertheless, an indefinite commitment, which is what that might well amount to, is unlikely, and funding has already been withdrawn once in the past.

It is questionable whether looking for micro fossils under extreme environmental conditions would ever be justified, bearing in mind 'difficulties' and 'costs', together with the fact that such life would presumably settle in less extreme conditions first, and we would be searching there anyway.

Nevertheless, we missed out on Extremophiles, one of the most unique and versatile of all known forms of life, through making exactly such assumptions, and we cannot afford do that more than once.

Meteorites will always be one of the commonest source of microbial samples, though contamination on entry is always a possibility, and only buried samples within the structure of the meteorite would give reliable information..

Terrestrial sites for evidence of extraterrestrial life will include any artifacts which suggest other than normal reasons for how or where they exists, and could include 'items' already scrutinize (e.g by von Daniken) if further doubts arise.

15.5 Alien Contact:

Assuming the existence of intelligent aliens, roughly as technologically as advanced as ourselves, there are three ways in which contact might take place - the first two passive (identification only, without interaction), while the third would involve face-to-face physical contact:

(a) SETI (11.8). This is the obvious initial choice, for which the necessary technology (radio telescopes) already exists. We have now been listening to selected radio frequencies for over 50 years, however, without success - an unexpected and disappointing outcome, which we hope can be explained by resolving Fermi's paradox. (15.6).

(b) Unmanned Probes: Two hypothetical probes have been postulated, by which aliens might explore our solar system in general term, or even in detail by surveillance from low earth orbit (or closer) though neither would be capable of soft landings or surface exploration.

The Bracewell Probe[8] is a self-supporting autonomous robotic probe, capable of collecting information, both physical and biological, with respect to indigenous life, and transmitting this back to it's own civilization. There would be no attempt at direct contact, but it would carry highly sophisticated software, with the flexibility to explore a wide range of situations.

Nevertheless, it's true usefulness is questionable, as is could take decades, or even centuries for information to get back to it's maker, by which time it would be out of date.

It was suggested at one time, that flying saucers might have been Bracewell probes, though all the anecdotal evidence suggests that they were manned.

Von Newman Probes[9] are an entirely different concept, designed to explore galaxies, preparatory to a civilization expanding outwards to eventually colonize that galaxy. Such probes are self-replicating space vehicles.

The initial probe travels to a nearby star system, where it obtains raw material from local sources (atmospheres, asteroids etc) to self replicate; each 'offspring' then travels to another star system, and repeats the process, with the total number of probes increasing exponentially, together with the number of star systems colonized.

It has been estimated that a galaxy the size of the Milky Way could be 'saturated' with probes in this way, in about half a million years. How long it would take to colonize thereafter, is less clear, but it would not be on an exponential basis, and could take many millions of years to complete.

However, over such a long period, unforseen advances in technology could change that scenario entirely, and colonization might be possible much sooner than that.

(c) Interstellar Travel:

We looked at some of the main options in chapter 13, and while time travel has undoubted advantages, it may still never come about, while the other options discussed, either depended on time travel, or on technology which we do not yet possess. There are alternatives, however, though no simple answers:

1. Conventional propulsion: for interstellar journeys, even the best technology we have would involve journey time of several hundred years, while devising a habitable vehicle would be almost as problematic as the journey itself.

2. Sub-light Velocities: It is common to denigrate these as impractical, but in fact at 10% speed of light (probably achievable with present technology) a journey to Proxima Centuri (the nearest star) could be completed in 40 years. In practice, however, there are no known suitable 'targets' within distances that would be realistic for this method of travel.

3. Hibernation: No form of suspended animation has yet been shown possible in human beings, and while computers could in theory monitor the journey, crews would nevertheless need to be available for emergency purposes, to supervise life-support and for the technical problems associated with eventual arrival.

4. Embryo transfer: Frozen embryos might survive the journey - but how they would be processed on arrival is a very different matter, and such a procedure could be impractical for the foreseeable future.

5. Time Dilatation. We looked at this in section 13.3, but it would be impractical from the point of view of reaching and sustaining the sub-light velocities required to bring it about, while the corresponding increase in mass would not be survivable anyway.

6. Life Extension. This was discussed in section 13.9, and probably shares similar difficulties in practice to those associated with hibernation.

15.6 The Silent Universe:

Fermi's paradox gave rise to a number of descriptive expressions, of which "Silent Universe" was particularly apt, and an instinctive response might have been, that we don't know about aliens because they don't exist. However, that would simply be stating the nub of the paradox in a different way, because on all grounds of logic and reason, they should exit!

Nevertheless, the casual way in which the comment was made takes a great deal for granted†. Much has happened over the 60 years since then, and we need to clarify a number of issues, to separate fact from supposition, before looking for possible solutions to the paradox itself:

FACTS:

- The universe is very old - 13,700 million years.
- It began in a Big Bang, and has been expanding ever since
- The sun is a young star - 4,600 million years old.
- There are billions of stars in the Milky Way galaxy, most of them red dwarfs, and most of these are older than the sun.
- Earth-sized planets orbiting within the habitable zones of G2 sun-like stars (and possibly red dwarfs too) are common.
- It took 9,000 million years before carbon, and other elements of life could build up to the necessary levels at which organic life could begin to form (7.1).

SUPPOSITION:

- Many of these earth-like planets could evolve life, and if so, most would progress to intelligent life.
- Whatever alternative life forms exist elsewhere in the universe, organic life will be a major part of it, and such life cannot be be significantly older (and therefore more advanced) than we are ourselves.
- Civilizations equally as advanced as us, are likely to rely on radio communication, in much the same way as we do.
- Advanced civilizations will eventually develop interstellar travel.
- A civilization capable of interstellar travel could colonize a galaxy, such as the Milky Way, in a few million years.

The PARADOX (strictly speaking an enigma - see preface) soon became an established tenet of modern cosmology, and though the words used were purely rhetorical, Fermi's colleagues apparently knew what they meant..

---

† "Where are they?"

Nevertheless, the implications are even more pertinent today, based on our greater knowledge, both of life, and of the universe itself, and in it's fullest sense the paradox embraces a number of features, both fact and supposition:

- Since we exist, life is a reality.
- Conditions and circumstances within the universe strongly favor the presence of organic life.
- These conditions are equally common elsewhere in the universe, and therefore it would be reasonable to expect to find organic life elsewhere in the universe.
- Some of that is likely to be at least as advanced as we are, and and therefore should be using radio communication very much in the same way that we do.
- With the large radio telescopes, and computer technology, developed over 50 years, we should have been able to pick up evidence of their radio activity long before now.
- Why has that not happened?

SOLUTIONS:

This paradox touches on a number of fundamental issue, and embraces a range of disciplines - astronomy, religion, philosophy, biology, ecology, among others. It has also generated a large number of scholarly papers and articles, and collectively over 50 different published solutions, yet not one has come up with a satisfactory explanation.

The difficulties relate not so much to the paradox itself, as to the balance between accepted fact, and the ambiguities of supposition, which some people regard as 'fact' (and weigh their argument accordingly), while others see them as 'unproven', and the arguments as invalid.

There is a lack of proper criteria, and too much flexibility of interpretation, and to try to avoid these difficulties we are adopting a somewhat simpler approach, based on the arguments presented so far, and the slightly more restricted criteria which these suggest as appropriate:

- The universe is the totality of everything which exists; nothing else is valid.
- organic life is the only form of existing life, and artificial life (as artificial intelligence) the only potential alternative 'life' form.

- Fine tuning' is universal.
- The chemicals of life began to build up about 9Gy after creation,.
- No form of life was possible before that.
- Evolution was overall uniform, and extinction events random.
- Hence no other life can be significantly older (more advanced) than human life.

Fermi's Paradox stems from just three words - these could be interpreted in many ways, but the clear intent was very general, and most people are familiar with the implications of Fermi's remark. That is one part of the overall problem. The other part is the paradox itself, 'why don't we know they exist', and an answer to that is the solution we are looking for, and which has been evading us for so long.

Unfortunately, our knowledge of the real universe, extensive though it may seem, is still very limited on the most basic issues - the 'Where', 'Why', 'When' and 'How' - of creation and life.

The paradox indirectly touches on all of these, and so there never can be a 'single' correct answer; but there should at least be 'consistent' answers, which do not conflict with each other, and collectively provide satisfactory reasons.

Just to be clear: the question at issue is not why SETI has been so unsuccessful, though it is often thought of in these terms, but rather why we are completely unaware of the existence of alien life elsewhere in the universe (given that we accept it must exist).

Many weird and bizarre explanations have been suggested in the past, most of which are readily accessible on the internet[10], but it is not the purpose of this book to elaborate on improbable proposals of that sort.

Rather, we have divided potential solutions into Three Main Categories, and the following discussion cover 20 possible explanations, which we consider at least plausible.

Those applicable to aliens themselves are inevitably hypothetical, but otherwise none conflict with known facts. We have intentionally avoided speculative proposals which fall outside of these simple guidelines:

1. ALIENS DO NOT EXIST, and terrestrial life is indeed the only life which has arisen (been created?) in the universe - so far.

This is the basis of the rare earth hypotheses, which is a complex statistical argument suggesting that the conditions for the existence of multicellular life are so complex, that they only exist together, in just the right proportions, on this one planet.

2. PROBLEMS OF COMMUNICATION - these do address the specific issue of SETI's lack of success:

- We have not been listening for long enough. Human life is only 200,000 years old, and man has only been emitting radio signals into space since the turn of last century - just over 100 years, or a tiny fraction of man's life span, and a totally insignificant amount of time in terms of cosmic events. If alien evolution was out of step with ours by even this tiny amount, the necessary match in technology would never have come about. In short, human beings may not have been looking for long enough, or alien civilizations might be that little bit too young, for this crucial overlap to have taken place at all.

- We have been monitoring the wrong frequencies. It is very difficult to know which frequencies to monitor, and the fact that aliens are aliens means that their radio technology, assuming they have it at all, may operate in very different ways from ours.

  The range of radio signals depends on signal energy, and that can be maximized by reducing the band width, which could then be picked up at much greater distances. The down side, however, is that the receiver must be tuned to that specific frequency, and also directed at that small part of the sky from which the signal is coming.

  The permutations are limitless, and while choice is based on what we regard as likely, that may not be the same for aliens.

- Missed signals. Those who have seen the original 'Wow' signal will know the sort of changes we are looking for, and it would not be difficult, if the signal content was minimal, to miss them altogether. However, over 50 years, that hardly seems likely

- Signals might be difficult to recognize. That would be much more likely. Again, the 'Wow' signal shows the pattern, and it is impossible to say what other form it might take, until we pick one up. One analogy suggested is comparing 'Times New Roman' with Egyptian hieroglyphics, as to how meaningful the content might be.

The digital world is probably fundamental to nature, but once we gets into 'compression', patterns lose their familiarity, and such signal might well go unnoticed.

- Radio signals may be infrequent or intermittent.. The 'quantity' and 'frequency' of radio activity, for any civilization, will vary considerably over the long term, and we can think in terms of three main phases: Discovery - both features are likely too vary significantly, until this new technology settles into it's permanent role, when use will likely be almost continuous. In the long term, however, as everything changes and advances, the use of radio may decline or eventually cease to be used.

SETI might simply have overlapped a period of irregular or minimal activity, especially towards the end, as use of radio was being phases out. That would certainly exceed 50 years, and there would likely be long periods of no activity at all, as it was finally being discontinued altogether.

- Classified Material! The potential uses of advanced alien technology would be almost limitless, especially military applications, and it may be that SETL was not only successful from the outset, but was able to established two-way communications.

From the moment that happened, the shutters would come down, with the risk of crucial information falling into the wrong hands so great, that SETI would have been instantly allocated the highest possible level of security.

As always, there are rumors that this has already happened - very similar to the NASA Apollo moon landings, which some have claimed never took place - but it would have been impossible to maintain this level of secrecy for so long - or even at all.

3. ALIEN REASONS why we may not know of their existence:

- Aliens may still be a primitive, non-technical culture. Given the size of the universe, and the number of likely alien civilizations, there might be a number of less advanced civilizations in proximity to us, and as noted above, as little as 100 year could make all the difference.

- Aliens may not use radio for communication. We only have to look at the ways in which communication takes place on earth, to see what the alternatives are. They may use fibre optics, for example, to a much greater extent than we do, while at a personal level, all aliens might be naturally telepathic. More highly developed sensory mechanisms, including auditory, may obviate the need to communicate by more conventional means, and the nature and extent to which radio is used may reflect these differences.

- The necessary technology may be too expensive. Funding has been described by one scientist as 'the root of all evil', and it may be just as important to aliens and their technology priorities, as it is to us.

- Aliens may find SETI intrusive and suspect our motives, An advanced civilization would have no difficulty in knowing that we were trying to listen in on them, and the concentrated and continuous nature of SETI might well suggest motives other than curiosity.

  Bering in mind that we are as alien to them as they are to us, such an interest could be alarming. Fear of an 'alien' invasion of their planet or even their galaxy, might be very real, and an immediate response would be to clamp down on all of their own radio activity.

  Beyond that, depending on how advanced they were, they might have ways to block such surveillance, or generate 'noise' which might distort or destroy the signals being received, while an even more advanced civilization might have the technology to render themselves invisible.

- Aliens might choose to ignore earth. There could be a number of reasons for that, from simple disinterest, to suspicion and fear of aliens, for their views on 'aliens' in general will no doubt be very much the same as ours.

- Aliens may already have self-destructed. This was, and many feel always will be, a very real risk for the future of the human race, and aliens may in effect, simply have fulfilled this prophesy.

  It is unlikely that destruction was total in a literal sense, but civilized society might have been all but obliterated, and whatever intelligent life remains, will be a primitive embryonic society, faced with the task of having to rebuilding it's whole future.

- Aliens may be totally dissimilar to us, anatomically and physiologically, as well as psychologically, and everything else about them and their culture may be equally different from ours. This would be reflected in how they behave, or even how they think and feel, and would be a major barrier to contact or communication of any sort. Something such as radio technology, or even the basic principles behind it, might have a different significance to beings such as these.

  If the biochemistry was also different, evolution itself would probably be very different from ours, and any comparison between the two might be meaningless.

- Aliens may be identical to humans. This is in marked contrast to the last answer, but if the universe is so finely tuned, 'carbon evolution' may be similar throughout. Negative SETI results could be accounted for in many different ways, while aliens themselves might be common, though their presence among us would go completely unnoticed.

- Deliberate avoidance of Contact. There are a number of possible reasons for this, some of which have been note already. An additional one might be if earth was an experimental project, or in some way special, and it's many unique features being monitored as a research project, with all forms of external influence scrupulously avoided. In this respect, some have likened earth's existence to that of a Zoo!

- Artificial Life. If one planet had reached a technological singularity (15.3), it would naturally seek to expand, at the expense of other life forms. This might be accomplished with suitable von Neumann probes (15.5), proliferating exponentially throughout a galaxy. Hence, rather than being the first site for organic life, earth might be one of few, or even the only planet, on which organic life still exists, and even that could be under threat.

- Non Temporal Concurrence. The huge size of the universe, and therefore light travel times, are so great that a civilization in some distant galaxy, might run it's course and cease to exist (for any number of reasons) before we even knew of it's existence. This effect could be even more wide spread than might be expected, due to the universe expanding.

- Privacy. The universe is enormous, and the consequences of different life forms 'meeting up' (by whatever means) so uncertain, that some inhabited regions (whether a single planet or a galaxy) might prefer to conceal their existence. This would require advanced technology, well beyond our foreseeable means, but might not be so for an advanced alien civilization.

15.7 Destiny - a hypothetical future:

Onward progress from the status quo at present, is an insidious combination of natural human progression and advancing technology, under the effective guidance and control of human intelligence. As thing progress, however, these three features will integrate and combine ever more closely with life itself, until eventually, evolution of what has hitherto been regarded as 'life', will slowly become that of something rather different - a functional combination of Machine, Intelligence and Man.

Outwardly, this will behave as a single functional entity, but internally the final conflict of existence is being resolved, as 'artificiality' confronts 'biology' in the battle for supremacy, which will determine the future from that point on.

Only that mythical 'higher intelligence' we hear so much about and see so little of, from the detached seclusion of an independent observer, will ever know the outcome

Evolution is a generic term which applies to all form of life, but it would be inappropriate to apply it to artificial life, while the functional triad, of which that is a part, simply 'progresses and improves' - whether one calls that evolution or not.

Intelligence can be thought of as an index of performance, and in that context has been suggested as a means of comparing humans with machines (Turing test), but it is a contentious issue, because what really distinguishes one from the other is not performs at all, but those intangible properties - morals, ethics, even a soul - and only our outside observer can ever know about these.

The essence of existence is 'presence', while the essence of evolution is 'change and improvement' - which are innate potentials in all forms of organic life, while machines cannot do either, and need to be recreated de novo in order to improve.

There is a certain irony in the fact that while human evolution is so slow, man nevertheless has the ability to create an exponentially evolving asset (technology) and use that to accelerate his own advancement.

The table in section 6.8 illustrates this, and also why as a consequence, we now live in a society totally dependant on technology, and would not only have been perceived as such by our independent observer, but also as a machine dominated existence in which man seems determined to build machines as clever as himself, in the ultimate opt-out from his human obligations.

Machines, however, have no concept of time, and can afford to wait indefinitely for the ultimate delegation - autonomy, the essence of which is independence from it's source, which would free man for ever from all obligations, though at the same time, would mean permanently renouncing his cherished status of 'dominant Species'.

By the time man realizes this, however, it is too late; all contact has already been severed, and his destiny changed for ever, from 'Principle and Director' to that of junior partner in a mechanical autocracy.

15.8 Conclusions:

The question of life existing elsewhere is not new. Philosophers have long taken it for granted, and though Fermi added topicality, most cosmologists probably already shared that view, but with the difference, that while philosophers can speculate, cosmologists must base their views on fact.

Fermi's comment (by all accounts 'off the cuff') was certainly appropriate, for 'brevity and mystique' are powerful incentives - in this case, not only for motivation (to start looking), but also expectations (of actually finding extraterrestrial life, from it's presumed use of radio communication).

Yet 60 years on (24/7 searching, 145,000 PCs and over 50+ explanations) no signals have been detected, and we are still not satisfied - for one very simple reason: SETI is not giving us the answers that we wanted, and which in fact it was set up to confirm in the first place (hopefully as a formality).

Nevertheless, there is nothing more destructive to objectivity than preconceptions and prejudice, and one thing is now certain, we are not going to find evidence of extraterrestrial life by continuing to listen in for radio signals. So how long do we go on looking, before accepting what SETI is clearly telling us - that our preconceptions were wrong in the first place?

Fermi's paradox is one of the few truly 'global uncertainties' in human life - whether we are alone in the universe, and whether earth is the only inhabited planet - and we are now in a position, for the first time, to draw conclusions based on fact. Whether these will be accepted or not, is another matter, and ultimately a choice between evidence or prejudice.

However, no decision (true or otherwise) can alter the eventual outcome, which depends only on existing reality, irrespective of what we thought or wanted that to be:

- From the facts presented in chapter 14, it seems unlikely that any alternative biochemistry could provide a substitute for organic life.
- Hence, Organic life is probably the only form of life.
- In addition, irrespective of earlier comments, it must be extremely unlikely that earth is the only inhabited planet, and therefore equally probable that organic life exists elsewhere in the universe.
- Advanced intelligent aliens, however, and the potentials attributed to them, while not impossible, more likely reflect wishful thinking than current reality.
- Moore's law has been shown to apply in other fields than electronics, and over a wide range of projects, progress is more likely to be exponential than linear; in other words, progress in technology is accelerating.
- The most important consequence of this is that artificiality (both life and intelligence) will continue to evolve ever more quickly, compared to the 'snails pace' of human evolution (6.8), and a technological singularity (15.3) is both inevitable and deceptively imminent.
- That will be the turning point for the future of life in the universe, marking both a permanent transition of species dominance, from organic to artificial, and the beginning of the decline of organic life.

- Man will first loose his veto on artificial activities, and then, as artificial intelligence progressively usurps organic prerogatives, and forges ahead, man by comparison will effectively regress - from subservience, through dispensability to redundancy! The pathos of all this will be entirely lost on man's new masters, emotions having been omitted from the original software, and in the very long term, organic life will simply be 'allowed' (perhaps even encouraged!) to die out, and artificial life will then become the only form of life throughout the universe.

- It may seem a bizarre scenario, but it is a logical possibility none the less, and we can then look back, from the vantage point of one of man's 'artificial progeny' at some infinitely remote future time, and with hindsight, see a universe filled with the beauties and accomplishments of an artificial reality, and at it's very outset, a tiny blip, in which some unknown form of life (organic?) brought such a future into being!

- And as for Fermi, who wanted to know "Where they all were"- a very different answer from what he probably expected, but a valid question nonetheless - just a bit premature!

- Speculation
August 2014

References (Chapter 15):

1. Are We Alone in the Universe (New York Times, Opinion Page, 18 November 2013).
2. Space.com 13 August 2013.
3. Wikipedia article (4 June 2014), Extraterrestrial Life.
4. Science Blogs, The Fascinating Logic of Cosmic Pluralism.
5. The Internet Encyclopedia of Science, Flammarion, Nicolas Camille (1842–1925).
6. J. Gorman, "Bacteria Found Deep Under Antarctic Ice, Scientists Say" (New York Times. 6 February 2003)
7. Wikipedia article (23 July 2014), Technological Singularity.
8. The Internet Encyclopedia of Science, Bracewell Probes.
9. M. Kaku, The von Newman Probe (big think, 18 Oct. 2012).
10. Google: 11 of the Weirdest Solutions to the Fermi Paradox (20 March 2013)

INDEX